The Day
the
MGM Grand
Hotel Burned

The Day
the
MGM Grand
Hotel Burned

by DEIRDRE COAKLEY
with HANK GREENSPUN, GARY C. GERARD,
and the staff of the Las Vegas SUN

LYLE STUART INC.   *Secaucus, N.J.*

*To those who died, in the hope that because of their
sacrifice, the world will be a safer place.*

First edition
Copyright © 1982 by Lyle Stuart Inc.
All rights reserved. No part of this book
may be reproduced in any form, except by
a newspaper or magazine reviewer who wishes
to quote brief passages in connection
with a review.

Queries regarding rights and permissions
should be addressed to: Lyle Stuart Inc.,
120 Enterprise Ave., Secaucus, N.J. 07094

Published by Lyle Stuart Inc.
Published simultaneously in Canada by
Musson Book Company,
A division of General Publishing Co. Limited
Don Mills, Ontario

Manufactured in the United States of America by
The Book Press, Brattleboro, Vt.

Library of Congress Cataloging in Publication Data

Coakley, Deirdre.
   The day the MGM Grand Hotel burned.

   1. Las Vegas—Hotels, motels, etc.—Fires and fire
prevention. 2. Las Vegas—Fire, 1980. 3. MGM Grand
Hotel. I. Greenspun, Hank, 1909- . II. Gerard,
Gary C. III. Title.
TH9445.H75C6   363.3′79   81-14511
ISBN 0-8184-0318-7   AACR2

Contents

OUR THANKS . . .

to Las Vegas *Sun* staff writers Bill Becker, Monica Caruso, Jim Coleman, Jeffrey German, Phil Havener, Harold Hyman, Elliot S. Krane, Penny Levin, Laura Lyon, Mary Manning, Mary Mele, Dan Newburn, Bob Palm, Chris Woodyard, and Scott Zamost, and to photographers Joe Buck, Ken Jones, Jim Laurie, Don Ploke, and David Lee Waite.

To Ardis Coffman, Irene Dunne, Captain Richard H. Kauffman, and Morag Veljkovic, for their valuable editorial contributions; to *Sun* art and graphics director Michael Matecko, librarians June Paramore and Barbara Rigby Connell, and typists Linda Azevedo, Vicki Broom, and Pat Johnson for their indispensable help.

And to Ruthe V. Deskin, assistant to the publisher, Las Vegas *Sun,* for her unfailingly fine assistance in seeing this project through from the beginning.

—Deirdre Coakley, Hank Greenspun, Gary C. Gerard

All the events set forth in this book are true; however, a few of the names have been changed.

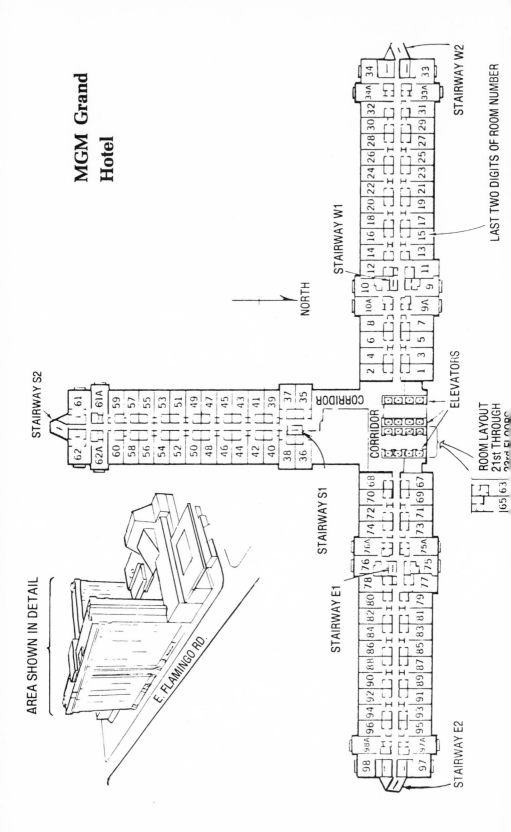

MGM Grand
Hotel

AREA SHOWN IN DETAIL

E. FLAMINGO RD.

NORTH

STAIRWAY S2

STAIRWAY S1

STAIRWAY E1

STAIRWAY E2

STAIRWAY W1

STAIRWAY W2

CORRIDOR

CORRIDOR

ELEVATORS

LAST TWO DIGITS OF ROOM NUMBER

ROOM LAYOUT
21st THROUGH
23rd FLOORS

While from a proud tower in the town,
Death looks gigantically down.

—Edgar Allan Poe
"The City in the Sea"

Prologue . . .

LAS VEGAS, *Friday, November 21, 1980.*

It was a few minutes past 7:00 A.M. on a sleepy Friday morning at one of the world's greatest hotels. The day shift was arriving to open the Deli, one of five restaurants grouped at the southeast end of the MGM Grand Hotel's cavernous casino.

Tile-maintenance supervisor Tim Connor was going about his rounds. He paused on his way to the kitchen when he heard crackling sounds and smelled smoke in a waitress station midway through the restaurant.

Connor spotted the flames of what appeared to be just another minor fire of the type that breaks out in kitchens across the land every day. He ran to notify guards of the small blaze centered in the kitchen wall.

Waitress Velma Turner, coming into the area from a different direction, saw flames near an electronic keno board in the restaurant. She hailed sous chef Ken Oborn.

Professional chefs and kitchen workers are knowledgeable about fires and are swift to react. Oborn grabbed a house phone and called the hotel switchboard, then ran into the kitchen and caught up a fire hose. As he aimed the hose at the blossoming flames, he shouted to the Deli's eighteen employees to evacuate. As they groped their way out through encroaching dense smoke, the small fire exploded with a *whoosh!* that knocked Oborn twenty-five feet out of the restaurant, into a bank of slot machines. He picked himself up and aimed the firehose back at the fire, but a second blast sent him sprawling again. Oborn fled.

Lights went out in the restaurant and kitchen, and in the sudden darkness, shouts of "Fire!" sounded through the area. It had been just sixty seconds since the flames were discovered.

The MGM Grand Hotel was ablaze. Within minutes, the news had circled the globe. Before the catastrophe ended, more than seven hundred persons were injured, and eighty-four dead.

This is what happened.

1 . . .

An Ordinary Day

Thursday, November 20, 1980.

It was an average kind of day in this community of four hundred and fifty thousand—but an average day in Las Vegas is unlike one in any other community. As always, there were conventions—a one-thousand-delegate meeting of the American Board of Clinical Immunology and Allergy was in progress at the MGM Grand Hotel, itself a mini-city, with its eight thousand guests and employees. At the sprawling Las Vegas Convention Center fifty million dollars' worth of computers and allied products were on display for the second annual Comdex Exposition. That event attracted some forty-five hundred exhibitors and visitors.

Evangelist Billy Graham was in town for a five-night crusade. His visit had started with a Governor's and Mayors' Prayer Breakfast at the MGM Grand two days earlier, and more than one thousand people had jammed the hotel's ballroom. The next night, backed by a chorus of a thousand voices, he had delivered his message to an audience of seventy-two hundred in the Convention Center's enormous East Hall.

At Nellis Air Force Base, ten miles northeast of the city, Tactical Air Command units from around the country were in the fourth day of Operation Red Flag, a month-long program of war games.

It was a blue and gold November day, with a high temperature of 67 degrees Fahrenheit—weather that sent many visitors to golf courses and tennis courts. For tourists just arriving from the chilly Northeast, heated hotel swimming pools beckoned. For many, the novelty of a late-November swim and a few hours of tanning were an irresistible lure.

The universal appeal of Las Vegas—the ambiance that attracts more than twelve million tourists a year to that city—has its roots in gambling, but there is more to it than that. For visitors, who began flocking to the desert resort when Beldon Katleman's El Rancho Vegas, the first small, rustic hotel and casino, opened on the Strip on April 3, 1941, Las Vegas has represented a special way of life.

The sprawling single-story El Rancho, destroyed by fire in 1960 and never rebuilt, led the way for an industry offering top-flight rooms, food, and big-name entertainment, along with exciting legal games of chance.

Luxurious hotels in fabled resorts around the world have always existed for the rich. Las Vegas was different. It opened up a whole new world of recreation for the average working person. As a score of other hotels appeared, each more fanciful than the ones before, the town became a place where the middle class and the wealthy mingled at the same gaming tables, sunned at the same turquoise swimming pools, and could see the superstars of the day, live and in person.

As a playground for adults, Las Vegas has no peer. As a fantasyland in the popular imagination, it long ago eclipsed the old dream city of Hollywood. But not entirely. In 1973, a resort was built that combined the best of old Hollywood and modern Las Vegas—the lavish MGM Grand Hotel. Twenty-six stories tall, with twenty-one hundred guest rooms and suites, the MGM Grand is a corporate descendant of Metro-Goldwyn-Mayer, the film studio that, during the long Golden Age of Hollywood, was the preeminent dream factory.

Las Vegas hotels have traditionally been theme-oriented. There is Caesars Palace, a Roman-Greco extravaganza, and the Aladdin, an Arabian Nights fantasy. Only in recent years has a crisp international style begun to transform resorts like the venerable Desert Inn, which began its life as a low, Bermuda-pink hotel. The MGM, starkly modern on the outside, was unabashedly red plush, gilt, mahogany, and crystal on the inside. Throughout the $106-million resort were huge blowups of the great MGM stars of yesterday. And the hotel's name, of course, derived from the 1932 Garbo hit *Grand Hotel*. For a population at the beginning of a nostalgia craze, the MGM Grand opened at just the right moment.

The MGM Grand achieved the hotel industry's highest daily occupancy rate, and by November 20, 1980, the hotel was 99 percent full. Guests were from every state of the union and from Japan, England, Ireland, Canada, France, West Germany, Australia, Spain, Italy, Argentina, and Mexico. The Mexicans—375 of them in large and small parties—had traveled to Las Vegas to observe the four-day national holiday celebrating the anniversary of the 1910 revolution.

By afternoon on November 20, the hotel's sixty-thousand-square-foot, 140-yard-long casino, brilliantly lit twenty-four hours a day with scores of immense crystal chandeliers, was humming with activity. Women in six-hundred-dollar Adolfo dresses were elbow to elbow with blue-jeaned players at the forty-five blackjack tables; high-rolling corporation executives pressed against the padded rims of ten dice tables alongside cowboy-booted package tourists. Bells rang and coins showered down in the trays of a thousand slot machines. For poker players, there were sixteen tables, and for true believers in the wagering of long shots, or just the weary of foot, there were two hundred comfortable chairs in the keno lounge and solicitous cocktail waitresses to provide complimentary drinks. Mute witnesses to the scene were dozens of life-sized marble statues, hand carved in Italy as classical grace notes for the luxurious casino.

It was to be an exciting opening night in the nine-hundred-seat Celebrity Room, where Mac Davis and comedian Lonnie Shorr were beginning what pre-opening publicity had heralded as "two fun-filled Thanksgiving holiday weeks."

At eight o'clock, in the packed nightclub, Lonnie Shorr opened the show; then Davis came onstage with a medley including "Stop and Smell the Roses," which he had written with Doc Severinsen, and several of his own songs, including his hit "Watchin' Scotty Grow." In a tribute to Elvis Presley, he sang a number he had written especially for the late superstar, "Something's Burning."

Of course, the Mac Davis concert was far from the only activity available to guests. Las Vegas resort hotels make it their business to offer something for everyone—enough distractions and facilities so that a guest need never go elsewhere, from check-in to check-out.

In no other place was this total-entertainment concept so much in evidence as at the MGM Grand. For example, hungry guests had a choice of five restaurants, including the pink-and-green Orleans Room, a twenty-four-hour coffee shop; Caruso's, for fine Italian food; and the luxurious Barrymore's, where diners were served Chateaubriand Bouquetier (thirty dollars for two), roast rack of lamb Diablotin (twenty-six dollars for two), or a variety of fish and seafood specialties, all under the eyes of John, Lionel, and Ethel, whose portraits looked down from the walls. Another kind of dining experience, considerably more expensive, could be had in Café Gigi, the exquisite gourmet room (Homard Grille Beurre Fondu, $27.50; Entrecote de Boeuf au Feu de Bois, $18.50; Oysters Rockefeller, $6.00). Dinner for two, to the accompaniment of Sasha Semenoff and his violins, could easily top $100.00, excluding the cost of ordering from the hotel's cellar of magnificent wines.

Something for everyone? Yes. The MGM also had a charming old-fashioned ice-cream parlor featuring every imaginable kind of high-calorie creation, and a New York–style restaurant called the Deli, with a high-Lindy's menu of lox, pastrami, and other savory specialties, such as the Cary Grant sandwich: Nova Scotia salmon, Swiss cheese, cream cheese, and Bermuda onion, $4.80.

If other attractions palled, there was the exciting jai-alai fronton, the first in the western United States.

And of course, the MGM Grand was the only hotel in town that could claim Clark Gable and Greta Garbo among its stars: Revivals of old Hollywood films danced on the screen of the three-hundred-seat MGM Theatre, where patrons sank into deep upholstered loveseats and could summon cocktail service by pressing a button. On that particular night, the old Gable hit *The Hucksters* was featured.

Nor were shopping opportunities overlooked. A grand staircase and escalators led from the casino down to the Arcade, a fantasy of forty specialty stores offering for sale furs, jewelry, antiques, designer dresses, avant-garde custom telephones, rare and beautiful silver-and-turquoise Indian jewelry, an art gallery, shoes, lingerie, candy, and, in keeping with the hotel's theme, the Nostalgia Shop. Visited by as many as twenty thousand browsers on an average weekend, the Nostalgia Shop stocked items that sold for as little as fifty cents and as much as twenty thousand dollars. Jack Benny's first violin fell into the latter category. And a customer once paid nine hundred and seventy-five dollars for the sandals worn by Charlton Heston in *Ben Hur.* Merv Griffin watched that transaction take place in the mid-seventies, and talked about it for five minutes on his television show.

Ironically, a replica of the *Titanic* was down in the basement, waiting for the December opening of producer Donn Arden's new seven-million-dollar extravaganza *Jubilee.*

Hallelujah Hollywood, the first show in the hotel's nine-hundred-seat Ziegfeld Theatre, ran seven years and played to five million people. That production closed October 12, 1980, and the Ziegfeld went dark in preparation for the new blockbuster. *Jubilee,* with a cast of 124, had been two years in the making. It would be in the grand tradition of *Hallelujah,* with scene after scene filled with beautiful women, toe-tapping music, and incredible special effects.

One production number centered around the sinking of the *Titanic,* and for this, a forty-foot replica of the doomed ship was built. Just three weeks before, to the amazement of motorists, the MGM's *Titanic* had been towed serenely along the Strip to the hotel. The three scenes of the *Titanic* number would also incorporate two fifteen-foot models of World War I fighter planes.

A Biblical show-stopper, Samson and Delilah, called for a thirty-five-foot-high statue of a sacred bull.

These vast, fanciful reenactments of disasters and mythical and historical events have always been a feature of the great Paris and Las Vegas stage spectaculars. The French Revolution, the London Blitz, and the crash of the *Hindenburg* were portrayed in earlier Donn Arden shows at the Stardust and Desert Inn hotels. In his production *Hello, Hollywood, Hello,* at the MGM Grand Hotel in Reno, Arden has a mockup of a Boeing 707 landing onstage.

For *Jubilee,* the show's 124 dancers, showgirls, and singers had been rehearsing for many months in the hotel's convention rooms and jai-alai fronton, as well as on the Ziegfeld Theatre stage when it was available. By November 20, after a late rehearsal, the show was far ahead of schedule. Most of the hard work was done, and all that remained were the lighting and various other technical aspects.

Arden and his longtime collaborator Margaret Kelly, the

legendary Miss Bluebell, whose dazzling Bluebell Girls have graced the stages of the world, are also responsible for the Paris and Las Vegas productions of the *Lido de Paris*. On that Thursday evening, they were jubilant over the state of their newest show, and Arden was heard to say that he was happier with *Jubilee* than with any other show in his life. He lives in Beverly Hills; Margaret Kelly's home is in Paris. During the exhausting months of rehearsals, they were staying in rooms on the seventh floor of the hotel. Among other out-of-towners staying in the MGM Grand for work on the production was Teresa Levitt from Los Angeles, assistant to costume designer Bob Mackie.

At the end of the last rehearsal, the keyed-up dancers— those "gypsies" immortalized by the show *A Chorus Line*— went out to celebrate the successful run-through of the production. Also keyed up, too happy to go to sleep, were choreographers Tom Hansen, Winston DeWitt Hemsley, and Rich Rizzo. Tired though they were, they found the thrill of seeing all their dance routines flawlessly executed, and in full costume, revitalizing.

"And such costumes!" they kept exclaiming to one another. But why wouldn't they be spectacular? The costumes, like every aspect of the production, had been done by the very best—by designers Mackie, Ray Aghayan, and Pete Menefee. The show featured eleven hundred of these stunning works of art.

Above the huge MGM Grand casino, that fantastic arena of green felt, crystal chandeliers, marble, and plush, gaming supervisors, seeing but unseen, took an occasional look down at the busy tables through one-way mirrors in the ceiling. These spot inspections of dealers and gamblers were made from a catwalk in an attic area among wires, pipes, and conduits. This "eye in the sky," intermittently patrolled, is a feature of every casino.

At midnight, the second Mac Davis show of the night was in progress. It was still early in the evening for many guests, and when the show ended just before 2:00 A.M., gaming tables were busy. Entertainment in the Cub Bar went on until 3:30; then, as the small hours of the night passed, play waned and table after table closed.

Outside the warm hotel, the temperature had dropped to 37 degrees Fahrenheit, and taxi drivers, police officers, and other outdoor night workers were feeling the chill.

Around the city, late-shift employees at hotels, hospitals, all-night markets, and fire and police stations were winding down. The last hour of night work is always the longest, and as the sun came up at 6:23 A.M., yawning graveyard-shift employees were ready to turn things over to incoming day people.

By 7:15 that Friday morning, the MGM Grand Hotel's ground-floor casino was nearly deserted. In the kitchen, cooks were readying orders, early-rising guests were calling for breakfast, and room-service waiters John Ashton and María Lucy Capetillo were starting upstairs with trays.

A handful of early risers sat in the coffee shop, scanning the morning newspaper over scrambled eggs and muffins or hot, buttered pancakes.

Several card dealers, in their crisp white shirts and black trousers, milled about the casino pit and exchanged small talk.

Security guards paced the casino and lobby in silence, the soles of their polished black shoes making no sound on the maroon-and-gold carpet.

Two bearded men drinking whiskey sours reminisced about the San Diego Chargers' twenty-seven-to-twenty-four overtime win over the Miami Dolphins the night before.

Janice Mitchell, a bookkeeper from Orange County, California, was nursing a glass of orange juice in the hotel's cozy piano-bar lounge. "Sue Ellen did it. I just know it," she said to her husband, who sat facing her at the small square table. That night, after months of suspense, viewers of television's hit show *Dallas* were to find out, once and for all, "Who shot J. R." TV fans around the world were waiting for the outcome.

Perry McGee, an insurance salesman from Tucson, Arizona, entertained himself at one of the casino's 970 slot machines. He had wrestled the tireless contraption's metal arm for almost two hours now, trying to win back the 180 dimes he had lost to the one-armed bandit.

Under the MGM's golden Grand Portal, Corey Brown, a computer programmer from Providence, Rhode Island, shut off the engine in his rented Ford Fairmont. He left the keys in the ignition for the hotel's valet-parking attendant and marched across the casino to breakfast.

Connie Mitchell was brewing coffee for employees who were struggling to hold back a morning yawn. The coffee pot in the coin cage had helped many casino workers make it through long winter nights.

At the registration desk, carefully groomed MGM Grand employees went about their morning routine. Guests departing on early flights would soon trickle down to check out. In an hour or so, there would be hundreds of people at the front desk, and by noon, the flood of outgoing and incoming guests might be counted in the thousands.

But in Las Vegas, unless the day's schedule calls for an early trip to the airport or a crack-of-dawn convention appearance, many visitors lie abed, catching up on lost sleep.

In a few hours, the action would normally pick up, and game after game would reopen for hundreds of players. But this early in the morning, only a few bored dealers ministered to the few gamblers. At many games, it was a one-on-one relationship. For example, at dealer Neal Barrett's blackjack table, he was dealing cards for a single customer. Barrett glanced off into the distance and saw a wisp of smoke drifting out from the doorway of the Deli.

A slight chill ran down Barrett's spine, but he didn't want to panic. He thought to himself, *We're here to entertain the guests, not to frighten them.*

Coin-cage workers and casino cashiers, nearing the end of an uneventful shift, counted the night's receipts.

In the casino change booth, cashier Alice Brown was balancing her money. Here, too, customers were few, and she looked forward to finishing a few minutes early and going home.

Relieved by Teresa Sperduti and Gigi Parrish, Alice was checked out by 7:15. She was on the hotel's lower level, headed for the double doors leading to the employee area, when she saw a short oriental woman dressed in the white kitchen uniform run through the hall.

"There's a bad fire in the Deli kitchen!" the woman yelled. Turning to a pair of MGM security guards nearby she cried, "You guys better get up there fast!"

Meanwhile, cocktail waitress Marla DuBois went about her rounds, with a business-as-usual smile. Kitchen fires are not that uncommon and are usually put out in no time at all. She considered dashing toward the kitchen, then thought, *Why tell them something they already know?*

A little woman dressed in white ran out of the Deli, waving her hands frantically in an effort to attract someone's attention. DuBois heard her utter the word *Fire!* and then watched as five security guards rushed to the east side of the casino.

At his slot machine, Perry McGee hit a cherry and two plums. Two dimes plinked into the tray. While feeding the machine another coin, he peered in the direction of the restaurants to see what the commotion was all about. It was just a little smoke, he thought. Nothing to get excited about.

The hotel's switchboard operator was calling the fire-alarm office in downtown Las Vegas. "There's a fire in the Deli kitchen here at the MGM. I think it's a grease fire. Send somebody."

At the other end of the line, dispatcher Judy Kusler bent over her microphone to call out the first-alarm response.

The mood inside Clark County Fire Station 11 was calm and relaxed that morning.

Fire Captain Rex Smith and the men who had staffed the

station since 8:00 A.M. the previous day were getting ready to head for home.

Some of them engaged in last-minute housekeeping chores as they awaited the men of the next shift, who were due to relieve them at 7:30. Others shuffled about in the station's modest recreation room, drinking coffee, talking, reading the morning papers, or playing pool.

In the garage sat a bright red fifteen-hundred-gallon American La France engine truck, its metal skin shining.

Parked in the dirt lot in front of the station, dwarfed by the colossal MGM Grand across the street, was Rescue Unit 11, the ambulance staffed by the station's two paramedics.

Except for the steady drone of people talking and the occasional thump of a polished ball on the billiard table, there was silence within the station.

Until 7:17. Then the speakers came to life: "We have a report of a fire in the Deli at the MGM." It was Kusler with her dispatch. "Station 11, Station 18, Engine 21 and Engine 12 respond. Utilize entrance number 2."

Smith and his men raced to the locker room for their turnout gear.

It would be a standard response. Station 11 would be the first crew to arrive at the scene.

Station 18, two miles away on Desert Inn Road near the Las Vegas Convention Center, would have its engine and rescue unit at the MGM Grand within barely more than 90 seconds.

Station 12, on Industrial Road behind the Stardust Hotel, and Station 21, on Tropicana Avenue about a half mile west of Interstate 15, each had one engine roaring toward the intersection of Flamingo Road and the Strip.

Engineer Chad Marshall revved the engine on the fire truck as Rex Smith, Toby Lamuraglia, Burt Sweeney and Ted Singer hopped on board. Paramedics Greg Redmond and Jeff Lovaas jumped into Rescue 11 and, with rooftop lights ablaze and sirens wailing, followed the engine across the street.

They pulled up alongside the hotel's north entrance—entrance number 2 in fire-department language—at the top of a ramp on the Flamingo Road side of the twenty-six-story hotel. Local residents call it the casino side door.

A light, yet steady stream of people poured out of the casino. With each quarter turn of the glass revolving door, Smith, a 13-year veteran of Clark County's firefighting force, noticed smoke puffing out into the brisk morning air.

There were no alarms, no bells, no screaming whistles, no sirens inside the hotel.

DuBois was still serving drinks to her customers, and some of them were making small jokes about the smoke, which grew thicker as the seconds passed.

"I don't know how long we can stay here if that smoke gets any worse," Barrett told the player at his 21 table.

"Deal," demanded the gambler, and Barrett slid two cards out of the shoe.

Apart from the clanging of coins at a nearby slot machine, there was still silence in the opulent casino. A shallow breeze blew across Barrett's face, and he heard the gentle tinkling of crystals in the giant chandelier above him.

The firefighters, met at the door by security guards, hurried across the hotel lobby toward the Deli. As they tramped through the hall in their heavy fire gear, the few guests at the registration desk glanced at the file of firemen, then returned their attention to the business of checking their bills.

Phyllis Thomas, graveyard-shift cashier in the coffee shop and soon to go off duty, was glad the firemen had arrived. The smoke was beginning to irritate her eyes.

She hadn't been scheduled to work that Thursday night, but her supervisor had called her to come in.

Suddenly the ceiling in front of the Deli erupted in flames, and the firefighters were chased back out the door. Shouts of "Fire!" and "Get Out!" filled the air, but at the far end of the casino, people sensed little immediate danger.

DuBois put down her tray and ran out the main entrance to the street, fully planning to go back inside when the fire was put out.

Barrett, thinking about the crush he'd be caught in if everyone started running toward the doors at the same time, swept the chips from his table and put them in the rack below.

"Come on—we'd better get out of here," he told his customer; then he dashed for the side door.

The gambler stepped away from the table, chips in hand, but did not seem to know which way to go.

"This way, sir!" Barrett yelled. "This way!" But the man seemed rooted to the spot, staring dumbfounded at the sudden commotion as the dealer fled.

McGee hit two oranges and a bell. Nothing. He deposited three dimes and yanked on the lever. Bar. Double bar. Plum. Nothing. Two more dimes. Cherry. Cherry. Bell. Good for one dollar.

Corey Brown, in the coffee shop next to the Deli, had taken his first sip of coffee and a bite out of an apple Danish when he saw the smoke. He threw down two one-dollar bills, a quarter, and a nickel and stormed out through the casino to the front door.

His car was still there, the keys in the ignition. The valet attendants had not yet parked it, and Brown jumped behind the wheel and burned rubber as he sped away.

Then came the holocaust.

As breakfasters, gamblers, and employees—screaming, running, stumbling—scrambled for exits, an enormous ball of fire rolled from the Deli along the ceiling into the hotel lobby.

2 . . .

Kerkorian's Empire

When the fire broke out, the MGM Grand's top executives—Al Benedict, Bernard Rothkopf, and Fred Benninger —were on a deep-sea-fishing boat off the Southern California coast, enjoying a brief respite from the pressures of managing a hotel-casino that was not only the largest, but also the most profitable, in the world.

In 1974, its first full year of operation, the hotel and casino returned the highest net profits and revenues in the fifty-year history of its parent corporation, Metro-Goldwyn-Mayer. The company's financial statement for that period revealed that of $28.6 million net income, the MGM Grand was responsible for $22 million. Not even in its heyday in the thirties and forties, when MGM studios were producing and releasing movies at the rate of one every week, did the money roll in so fast. (And during that heyday, the movies that came out of MGM were those which most Hollywood historians point to as the industry's greatest. They included *The Wizard of Oz, Mutiny on the Bounty,* the Gable, Garland, Garbo, Tracy, Hepburn, and Crawford pictures, the Busby Berkeley musicals, and the wartime dramas.)

It was no accident that the Hollywood film company planted itself foursquare in the booming Las Vegas resort market. Kirk Kerkorian, majority stockholder of the film corporation since 1971, made it all happen.

Even in a town of legendary entrepreneurs, Kerkorian is an original. "My God, who wants to read that stuff over and over again—how I sold papers as a kid and made good?" he remarked to an interviewer in 1969. But the truth is that his progress to a position in the highest regions of international finance was remarkable.

Born in Fresno, in California's central valley, to an immigrant Armenian couple, Kerkorian came of age during the Depression. He did indeed hustle newspapers. Later he worked in a CCC camp; then he bought used cars and resold them.

Wiry and athletic, he thought for a time of a boxing career. He was considered a very good boxer and might have had a better-than-average chance of succeeding in the ring, but his imagination was captured by the idea of learning to fly. That was a fairly adventurous notion in pre–World War II days, but he took lessons and became an accomplished flyer. When World War II created an instant need for pilots, Kerkorian was put to work as a flight instructor for the U.S. army. After a time, he switched to ferrying planes to England for the RAF.

In the airplanes of the forties, flying the Atlantic was

tough, demanding, risky duty, but he emerged from wartime service uninjured, a highly seasoned pilot.

Along with many other young men who had served in the air, he saw peacetime commercial aviation as a promising field, and when the U.S. government began offering surplus aircraft for sale at rock-bottom prices, he recognized an opportunity. World War II planes glutted the market, and cheap as they were in the United States, they were even cheaper in the territory of Hawaii. Putting together every dollar of cash and credit he could summon, he made purchases in the islands and ferried the planes to the mainland himself, often at considerable risk.

On one flight, a DC-3 he was piloting home developed engine trouble and threatened to go down midway across the Pacific. He nursed the stricken plane to a safe landing in California; after touchdown, he was said to have coolly shrugged off the near ditching.

Those surplus planes were the nucleus of a small supplemental cargo and passenger airline he started, one of many such hopeful ventures launched by former wartime pilots. That most of his competitors went broke, while Kerkorian's fledgling business grew into the giant Trans International Airlines, has been attributed to his unusual ability to foresee coming trends. He was the first supplemental-airline owner to buy a jet when everyone else was still flying piston planes, and that jet made the difference.

Kerkorian was a frequent visitor to Las Vegas, and in the sixties, he acquired land on the northwest corner of the Strip and Flamingo Road which would be the site of Caesars Palace. He was that spectacular resort's landlord for a time, and in later years, when he sold TIA to the giant

Transamerica Corp., the millions he received for his stock became the foundation for his initial entry into hotel ownership in the burgeoning Las Vegas resort market, first with the aging Flamingo Hotel, which he purchased for fifteen million dollars in cash in 1967, then with the International, a Kerkorian enterprise from the ground up. Other acquisitions in the early seventies included controlling interests in Western Airlines and in Metro-Goldwyn-Mayer.

When Kerkorian's fifteen-hundred-room International Hotel opened on the July 4th weekend in 1969, it was then the world's largest resort hotel. The luxurious, thirty-story hotel-casino, adjacent to the Convention Center, brought forth all the superlatives in the vocabulary of grand openings. The glittering hotel was aswarm with VIPs and celebrities, and every facet of the resort was faithfully recorded by a hundred travel, sports, and entertainment writers flown in for the event.

Barbra Streisand was the grand-opening star in the main showroom, and Peggy Lee, a big-room star in any other hotel, appeared in the lounge. The International, two and a half years in the making from first announcement to completion, set new records in many respects. As contractors on the giant structure raced to make the grand-opening deadline, many guest rooms were still unfinished. While the Beautiful People roamed through the ground floor waiting to be conducted to their rooms, those rooms were still getting last coats of paint; as hotel workers moved in furniture and made up the beds floor by floor, word was relayed to the registration desk as each room was ready for occupancy. It was that kind of an opening, but everyone had a wonderful time and the new mega-hotel was pronounced an overwhelming success.

Kerkorian sold the International and the Flamingo to the Hilton chain a year later, then set about to create yet another gigantic resort, one even larger and more lavish. This was to become the MGM Grand Hotel, and to finance its construction, he sold off many assets of the movie studio, a move that infuriated both stockholders and Hollywood sentimentalists. His rationale: to put the corporation heavily into the booming leisuretime industry, since moviemaking was no longer predictably profitable.

His expectations were borne out by the enormous success of the hotel. A second, hugely popular MGM Grand made its debut in Reno, 450 miles to the north, in 1978, and in 1980 a third, in Atlantic City, was in the planning stage. At the flagship hotel in Las Vegas, construction of a 782-room addition was underway. With no interruption of daily business, crews labored early and late on a new tower going up on the southwest corner of the building.

The MGM Grand was colossal, a glittering total-entertainment complex designed to outdo anything else in the highly competitive Las Vegas resort industry.

Called upon to create it were the same companies used for the mighty International—among them, the architectural firm of Martin Stern, Jr., and Associates of Beverly Hills for design, and Taylor International Corp. of Las Vegas, which had also built the Fontainebleau and Eden Roc in Miami and Caesars Palace, the Riviera, and the Tropicana on the Las Vegas Strip. As managing contractor, Taylor International would oversee all phases of construction and furnishings, from groundbreaking to completion.

When the MGM Grand in Las Vegas was planned, the site for the new hotel was a forty-three-acre parcel at the southeast corner of the Strip and Flamingo Road. The location was partially occupied by the ill-fated 160-room Bonanza Hotel, which had opened for business in July 1967. After a series of financial reverses, the rambling, western-style hotel-casino had closed in October of that same year. It was subsequently bought, sold, then repurchased by Kerkorian. So the old Bonanza property, plus additional land acquired by the financier, formed the site of the new MGM Grand Hotel.

It was no ordinary groundbreaking, just as the MGM Grand would be no ordinary hotel. A giant tent was erected in the desert along the Flamingo Road side of the property, and on April 15, 1972, at a champagne-and-caviar party attended by Cary Grant, Chad Everett, Janet Leigh, Betty Grable, and scores of other stars, celebrities, VIPs, and members of the press, the beautiful Raquel Welch depressed the plunger to set off a charge of dynamite, signaling the beginning of construction. Hundreds of balloons of every color were released into the air, and a barrage of fireworks forming a giant picture of Leo the Lion burst across the sky.

The bright tent had barely been dismantled and trucked away before construction crews were hard at work.

As building progressed, more than a hundred subcontractors were involved in the project—all of them impressed with the need for haste. By any standard, the MGM Grand was a rush job. The two-and-a-half-million-square-foot hotel opened nineteen months after groundbreaking and less than sixteen months after the first steel was put in place. Even so, it was 210 days ahead of schedule. The rapid completion was credited to "fast-tracking," a process in which a building is under construction even before final plans are drawn; in fact, the subcontractors frequently drew their own plans for the building's components.

Six years later, Taylor International president Stuart J. Mason would explain fast-tracking in court during the trial of a hundred-million-dollar damage suit filed by the MGM Grand against Johns-Manville, one of those subcontractors. In that suit, in which a Clark County jury awarded damages of $14.6 million, MGM attorneys charged that the original design for the hotel called for a stucco exterior on the walls but that Rescon, a gypsum wallboard, had been substituted. "The product had never before been used on a high-rise structure," the MGM management charged, "nor had it been tested to withstand heat, wind and moisture. . . . As a result, several of the 4-by-10-foot boards or panels have fallen. Imagine what would happen if they hit someone!" According to witnesses, the insides of the panels calcinated—just turned to powder—in the extreme Las Vegas heat.

According to Mason's testimony in the landmark case, "We don't tell [the sub-contractors] how to get there necessarily. Like on air conditioning, we would say: 'We want this room to maintain 75 degrees, and the room next to it 72 degrees.' We list all the rooms . . . and it is up to them to design the type of system that is going to achieve what we want." Thus the MGM Grand, what the *Los Angeles Times* called "a hotel in a hurry," was the joint effort of many designers.

To bring the whole project to completion, as many as a thousand workers swarmed over the site. On the ground floor, in the immense area that would become the hotel's giant casino, job supervisors drove from point to point in pickup trucks.

Some of the statistics: The job required 15 million square feet of gypsum on interior walls, ceilings, soffits, and pillars; also needed were 11 acres of roofing; 450,000 square feet of waterproofing material; 750,000 square feet of stucco for exterior walls, and 300,000 square feet of flame-retarding material and insulation on cooling ducts and water piping.

There were 8 acres of glass, including 55,000 square feet of reflecting glass, 142,000 square feet of one-way glass, and 160,000 square feet of window panes tinted gray or bronze.

The building also called for 20,000 tons of structural steel and 1,895 miles of electrical wiring—enough to reach from Las Vegas to Detroit. Fitting out the structure required 50,000 lighting fixtures, 200,000 light bulbs, and 2,300 television sets. The hotel had 12 million watts of heating and cooling capacity, enough to service a town of 8,000.

Construction was not without its problems. On July 17, 1973, when the project was less than six months from completion, the more than 100 members of Ironworkers Local 433 at work on the hotel tower walked off the job in a protest over safety conditions.

"We've tried every way we know how to get this job made a safe one," union steward Jack Herron said to the men. "Now we're going to sit on our duffs and *pray* for safety; that looks like the only way we'll get it."

Fred Toomey, business agent for the union, explained, "The railings aren't properly secured, and they aren't all the right height—if you tripped, they wouldn't stop you from falling off. And there's a lot of trash up there that should be cleaned away, because it's a fire hazard with the welding going on. We've been unable to get it cleaned up by complaining to the state inspectors." Cleanup and safety devices, he pointed out, were the responsibility of the general contractor. The subcontractors and their employees were paid only for the ironwork itself. The dispute was resolved and work was resumed when Taylor Construction agreed to pay the men to correct the safety hazards.

On December 23, 1973, as searchlights played across the starry Las Vegas sky in the grand tradition of a Hollywood premiere, the MGM Grand opened its doors with a lavish VIP party. As mentioned before, opening-night star in the Celebrity room was Dean Martin. Before and after the show, invited guests toured the vast structure, previewing some of its thousands of wonders, including the Gone with the Wind Suite, a two-story fantasy out of a Hollywood set director's dreams; the Metro Club, an exclusive penthouse casino for the highest of high rollers; and guest rooms, replete with movie-star bathrooms and a gold star on every door along the miles of corridors.

Later, designer Donald Schmitt, who did the interiors of the International, the MGM Grand in Las Vegas, and then the MGM Grand in Reno, would say, "The important thing is to give the gambler the feeling that an unlimited amount of money has been spent on him. The trick of good design is to be flamboyant but not cheap. And actually, that requires not so much a trick, but the use of the best materials. You've got to use the real thing."

The MGM Grand opened on December 24, 1973, and seven years later, the hotel in a hurry, a smooth-running, computerized gold mine, was internationally acclaimed. Kerkorian, as was his habit, was a distantly perceived presence. Other interests in 1980, including a 24 percent ownership in Columbia Pictures, were commanding his attention. Day-to-day management of his properties had never been his style. For this, he had the best people, and he paid well to get them.

The many similarities between Kerkorian and the late Howard Hughes were never overlooked by observers, but although both men were involved in flying, airlines, movie-making, and Las Vegas hotels at various points of their lives, there were significant differences in their careers and methods of operation. First, of course, was the fact that Hughes started rich, and got richer. Kerkorian, with only an eighth-grade education but astonishing business ability, started from scratch.

Were Kerkorian and Hughes rivals in Las Vegas? In 1969, with the opening of the International, Hughes's properties, the Desert Inn, the Sands, the Frontier, the Castaways, and the Landmark, were actually eclipsed in total room count by Kerkorian's two hotels.

Hughes was perhaps the world's most famous recluse; Kerkorian, although he generally shunned the spotlight and deferred to his chief of staff, Fred Benninger, as corporate representative, maintains a certain amount of informal visibility, playing tennis, yachting, smiling shyly at such events as the 1973 groundbreaking of his MGM Grand Hotel.

Kerkorian apparently believes in delegating authority; Hughes, until his last sick, agonizing years of exile around the world, frequently wanted control over some of the smallest matters regarding his properties. His top executives fumed while they awaited word of his decisions on what would have been routine middle-management judgments in any other corporation.

But one trait Hughes and Kerkorian shared was a recognition of the necessity to employ the best available people. Often they employed the same people—for example, Al Benedict and Bernard Rothkopf, both former Hughes men, joined Kerkorian's organization when the MGM Grand was in the early stages of planning.

Benedict, MGM Grand Hotels president, was born in Philadelphia. He served six years in the merchant marine, then attended Rutgers University, where he was a swimming champion and a member of the 1947–48 All-American Swimming Team. In 1952, he began his hotel career with the old Last Frontier in Las Vegas. He later served with the Desert Inn, then was general manager of the Stardust when he was tapped for the job of director of Hughes hotel operations. A policy clash with Hughes Tool Co. executives led to his leaving the organization in 1971. Kerkorian snapped him up. At the MGM Grand Hotels, he presided over two of the world's most profitable resort hotel complexes.

The second member of the triumvirate running the MGM Grand in November 1980 was Cleveland-born Bernard

Rothkopf, who arrived in Las Vegas in 1950 as a founding partner in Wilbur Clark's Desert Inn. In the mid-fifties, Rothkopf was instrumental in opening the Stardust and Showboat hotels. A man of excellent experience and sterling reputation in gaming, he was made executive director of the Desert Inn by Howard Hughes, then transferred to the ailing Sands Hotel to work his magic.

He joined Kerkorian's MGM Grand team when the Las Vegas hotel was under construction as executive vice-president in charge of gaming; later, he was made president of the hotel.

The third member of the MGM Grand's top management team was Fred Benninger. In 1967 Kerkorian enlisted Benninger to be chief administrator of the entire Kerkorian financial empire. A *cum laude* graduate of New York University and the University of Southern California, Benninger began his business career before World War II with the international accounting firm of Arthur Anderson and Co. He moved on to Flying Tiger Lines in 1946, and twenty-one years later, when he left to join Kerkorian, he was Flying Tiger's executive vice-president.

In the all-important area of gaming, veteran casino executive Carl Cohen, formerly vice-president of all Hughes hotel-casino operations, was executive vice-president of the MGM Grand Hotel's casino operations, and Morrie Jaeger, for many years a Caesars Palace gaming executive, was vice-president in charge of gaming at the MGM Grand, Las Vegas.

Others in the organization, from entertainment chief Bill DeAngelis, to food and beverage director Pierre Vireday, to ballet mistress Ffolliott (Fluff) LeCoque, to head engineer Scotty Bell and on through the hotel's myriad departments staffed by forty-five hundred employees, were among the world's most seasoned, highly regarded professionals in the difficult business of providing luxurious accommodations, food, drink, entertainment, gambling, and services to 1.5 million guests a year.

On that fateful Friday morning, the fishing boat carrying Benedict, Benninger, and Rothkopf bobbed in the coastal waters as the rising sun lit up the sky. The men, looking forward to a full day's fishing, had every reason to believe that all was well back at the hotel.

3 . . .

The Fire

By 7:30 Friday morning, a deadly blaze was raging through the Las Vegas hotel. The devastating fire, inexorable as a tidal wave, had begun its rampage.

Outside, Fire Captain Rex Smith radioed for additional units, and Clark County Battalion Chief Al Hurtado immediately declared the fire an evolution three—a life-threatening blaze.

Fire department units from Las Vegas, North Las Vegas, Henderson, Boulder City, Nellis Air Force Base, and the rest of Clark County pulled out of their stations and raced toward the MGM Grand. Inside the casino, crystal chandeliers crashed down and ceiling panels cracked and fell, while above, so far unseen by anyone, a second fireball raced toward the front entrance.

Pipes, conduits, and insulation material ignited and burned with no letup. A moment after the wall of fire had crashed through the restaurant end of the casino, a second,

40

more violent fire reached the hotel's main entrance, dropping a blistering wall of flames that consumed everything in its path. Carpeting, gaming tables, light fixtures, furniture, ornaments, and wall coverings were all engulfed in seconds.

The plastic faces of slot machines melted into viscous pools of toxic resin. Paper money, tossed into the air by the fire's turbulence, was carbonized. Plastic chips, upholstered chairs, and stools were transformed into distorted, other-worldly shapes. As if in a replay of the last days of Pompeii, classic marble statues lost noses and ears.

It was a vast, impersonal force. An incredible one million cubic feet of fire, at twenty-four hundred degrees Fahrenheit, roared through the casino and out the hotel's main entrance, where an unoccupied automobile parked under the Grand Portal was incinerated in moments, as hundreds of fluorescent lights popped from their sockets, showering the driveway under the flaming canopy with blackened cylinders.

In just ninety seconds, life ceased in the casino and surrounding areas. Fourteen persons were already dead.

As the fire raged, a cloud of killer smoke poured upward through the building.

Laden with carbon monoxide and cyanide gas released by the burning plastics, the smoke traveled through every crack and crevice, through every duct, through every open door, rising through elevator shafts and ventilators, blowing into stairwells, into corridors, into luxury suites and guest rooms, even creeping through bathtub drains.

Escaping into the open air, the smoke forced a column that, in a matter of moments, towered nearly a mile.

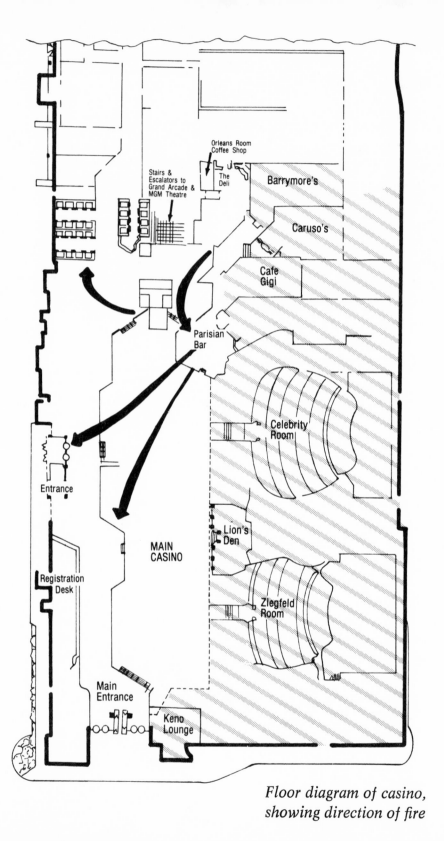

Floor diagram of casino, showing direction of fire

Clark County Fire Chief Roy Parrish, in his radio car on the way to work, heard Hurtado's evolution-three call. He had helped fight the El Rancho Vegas fire two decades before. Racing to the scene, his car's "redheads" flashing, he turned over in his mind the logistics of how to fight this one.

At the hotel, coin-cage cashier Alice Brown was directing traffic. She had continued on her way out that morning, undisturbed by the Oriental kitchen worker's cry of "Fire!"

She had punched her time card, exchanged small talk with some fellow employees, and headed across the parking lot toward her car.

When she took a casual look back, she saw a wisp of smoke rising from the roof. A moment later, the whole roof was billowing forth thick black smoke.

Day-shift workers started arriving, and traffic on Audrie Street, the narrow artery linking the employee parking lot with Harmon Avenue, picked up.

Brown, a Nevada Emergency Teams (NET) director and registered emergency medical technician, didn't want the street blocked, because she knew the fire engines would soon be roaring in. She began waving her arms to warn cars away, and within seconds, the wailing of fire sirens filled her ears.

The smoke slithered up the highrise like running water. It was less than ten minutes since she left the cage. She had smelled no smoke.

Nearing the hotel, Parrish knew from the smoke that it was going to be bad. By 7:25, when Parrish arrived, the second wall of fire had crashed through the main entrance. The casino was already a raging furnace.

In his mind, as he watched what was happening, he thought, *I hope no other fire chief ever has to see what I'm looking at this morning.*

Meanwhile, Fire Captain Smith recalled his bright red engine from the standpipe down the street and radioed for a hosewagon. Then, collapsing the revolving doors leading to the casino and lugging a swelling fire hose, he led his men back inside.

As the struggle to subdue the flames began in earnest firefighting equipment, some from as far as seventy miles away, continued to race toward the anvil-shaped cloud of smoke.

An engine pulled up near the hotel's Grand Portal just as the wall of flames crashed down. The firemen had planned to rig a hoseline and enter the casino lobby through the main entrance. But now there was no way in.

Fire Chief Parrish, after looking at the hundreds of guests huddled by the windows, thought, *We've got to control the fire and keep it out of the high-rise.*

If the fire spread to the upper floors, Parrish knew, the devastation would be beyond imagining. Effective firefighting is difficult—if not impossible—beyond the ninth story of a building. Fire crews have to climb up the stairs, carrying all equipment with them—that is, if they have the time. Given the nature of the MGM fire, it would be too late. The blaze would spread up through the elevator shafts as had occurred with the flash fire in the casino, and the flames would have ripped through the corridors, consuming wallpaper, carpeting, and anything else in the way. Because of this flashover effect, the fire could conceivably race through the corridors in only a minute or two. Thousands of people, trapped in the high-rise, could be lost.

Glass and metal fragments rained down on the firefighters entering the casino from the north entrance. Smith and his men worked two-and-a-half-inch hoses back and forth over the flames. They fought their way to the elevator area, where the heat, estimated at twenty-eight hundred degrees Fahrenheit, had already melted the elevator cables, sending two cars crashing to the basement. To avoid another deadly flash fire, Smith knew, they had to keep the heat from building up again in the shaft. The three fire crews inside, each made up of ten to twelve men, concentrated on keeping the elevator area doused. Often, the fire would whip around behind the men, blocking their way out. Then it became a matter of fighting the flames just to stay alive.

On every floor of the T-shaped hotel, guests waited behind smashed-out windows and on balconies. With smoke billowing out from a thousand directions, they waved bright yellow bedsheets to signal for help.

Firefighters swung ladders as high as they could. The longest ladder used by the Clark County and Las Vegas fire departments—and by most departments in the United States—is 100 feet. A nine-story reach was all that was possible.

From that height a handful of guests started down, skimpily clothed, some bleeding from broken glass. Firefighters encouraged and steadied them as they descended.

More than 250 men had a hand in the fire departments' efforts that morning.

Volunteer units, called in by county firefighter Richard

Wiseman, responded from the rural communities of Moapa, Overton, Goodsprings, Mountain Springs, and Blue Diamond.

The Boulder City Fire Department sent two rescue units. The special-operations wing of the 20th squadron from Luke Air Force Base and the 57th Fighter Wing from Indian Springs dispatched a total of nine Tactical Air Command helicopters to the scene.

Nellis Air Force Base firefighters were put on standby at Clark County Fire Station 16, on Nellis Boulevard, to handle a potential emergency that might develop in the Las Vegas Valley. Fortunately, there were no other major fire or rescue calls.

One man involved in the rescue effort was a firefighter from New York City. An MGM Grand guest, he escaped from the hotel and raced straight for a firetruck. He identified himself to a county fireman, grabbed a coat and boots, and went to the aid of hotel guests who were being led out of the building.

Shortly before 6:00 A.M., uniformed and plainclothes Las Vegas Metropolitan Police Department officers had served warrants that resulted in the arrest of thirty-three people in connection with a major narcotics trafficking operation.

On a normal day, the news of the arrests would have been prominently displayed on Page 1 of the three Las Vegas newspapers. But this Friday was far from normal.

Metro Police's entire graveyard shift was held over and much of the oncoming day shift was dispatched to the

MGM Grand. As the police helicopters, Air 1 and Air 2, took to the air, K-9 officers and their German shepherd partners set up barricades in an effort to hold back the growing throng of onlookers.

Soon firefighting efforts were well underway. Every twenty to thirty minutes, as the yellow tanks strapped to the firefighters' backs ran out of air, the men emerged from the inferno to change tanks. Captain Smith said that by his third air tank, "We had succeeded in knocking the fire down in some areas. The men were getting worn out because of the heavy hose line and the great concentration of heat."

The new shift replaced them.

By that time, Chief Parrish had assigned Battalion Chief Al Hurtado and Assistant Chief Bob Atkinson to coordinate the firefighting effort. Battalion Chief Paul Hicks was put in charge of the rescue operation.

As firefighting and rescues continued, Metro Police helicopters Air 1 and Air 2, equipped with bullhorns, started flybys around the building.

"We told people, 'The fire is not close to you. You will be all right,' " Parrish said. "We had a language problem at first, but we solved that. The department recently hired a mechanic who speaks German fluently, and we commandeered a citizen out of the crowd to talk to the Spanish-speaking people."

A heavyset man started climbing down some sheets he had tied together.

"When I saw that first guy come down that rope, it

scared me," Parrish recalled. "He fell and came down hard." The man was rushed to the hospital with serious internal injuries. "I had a very eerie feeling that other people would try it."

One who did was Randy Howard, a termite inspector from Moline, Illinois, who had come to town to gamble and to see Ann-Margret, who was playing at Caesars Palace. Awakened that morning by pounding on his door on the fourteenth floor, he thought it was "some drunks, screaming and running down the hall."

Then he smelled it. In the corridor and outside his window, all he could see was a wall of black smoke. "It was so thick it was like we were in a chimney," he said of the floor where his room was. "People were starting to collapse. The smoke was killing them. They didn't want to move. They just sat there tranquil.

"I wanted to lay down and sleep. I knew I was ready to pass out. I dropped to my knees and yelled, 'God in Jesus, help me find a way out!' Just then I looked out the window and saw a rope swing by."

He left his room and ran down the hall, trying to find a room from whose window he could reach the rope. In the black hallway, he bumped into someone every few feet. "I ran into two ladies who said, 'We're going to die, aren't we?' Death was there, and they knew it."

Finally, he found the room where the swinging rope had stopped. Although Howard did not know it, the rope had been dropped by firemen who were preparing a boatswain's chair to lower people to the ground. He saw it as his only chance to escape.

"I started going down hand over hand," he said, "fast

but not allowing myself to slip. When I got four flights down I thought I was going to pass out for sure. I felt that if I dropped only eight or nine flights, I might live."

His cowboy hat saved his life. From the floors above, falling glass broke on the hat.

"Just when I thought I was blacking out," he said, "my feet hit something solid. My foot was on the arm of the man who fell."

Reporters, who had watched his perilous descent, approached him as he ran into the street with glass exploding behind him.

He showed them his hands, which had not a single rope burn, despite his fourteen-story descent. "I'll go to my grave believing that this was an act of God," he said.

4 . . .
Media Coverage

Five television stations, a score of radio stations, and three daily newspapers serve Las Vegas. Of the newspapers, the *Las Vegas Sun,* called "the most influential newspaper in Nevada" by the *Wall Street Journal,* was founded in 1950 by Hank Greenspun. Then, on November 20, 1963, a pre-dawn blaze swept through the newspaper plant, destroying everything in its path.

Hank Greenspun rebuilt his newspaper, and now, on a morning seventeen years and one day later, on the twenty-eighth floor of his highrise condominium two miles from the MGM, Greenspun watched, with mounting horror, the incredible sight of the MGM Grand Hotel erupting in flames.

"Barbara," he shouted down to his wife. "Bring me my binoculars!"

"Hank, come down from there before you break your neck," she called back.

She ran to the apartment of a neighbor, whose windows faced the MGM Grand, and asked if they could watch from their balcony.

The neighbors, Eugene and Gerrie Maday, owners of several stores in the hotel's lavish Arcade, invited the Greenspuns in.

Maday, appalled at the sight of the fire, mourned: "It's all gone . . . it's all gone." He would be one of the many shop owners put out of business that day.

As the Madays contemplated the scene, Greenspun, on the telephone with the *Sun* offices, watched, with mounting horror, the incredible sight of the MGM Grand Hotel erupting in flames.

At seven o'clock on Friday, November 21, the newsroom of the *Las Vegas Sun* was nearly empty.

Assistant managing editor Jim Barrows and copy editors Cristine Soliz and Russell Bert labored silently at their computer terminals, preparing stories for that afternoon's Update edition. They were alone in the room of empty desks and silent telephones.

Suddenly, the fifty-channel Bearcat police scanner squawked to life. "We have a 402 at the MGM." An instant later, police reporter Gary Gerard was on the phone with Len Butcher, who had stepped down as the *Sun's* managing editor the day before.

"Get a photog out to the MGM," Gerard said.

"David's already on his way," Butcher countered. David Lee Waite is the *Sun's* photo editor.

Gerard, who could see the flames at the hotel from his tenth-story apartment window, described the scene over the telephone: "Flames are shooting pretty high over the front canopy. There's a ton of smoke. It's the worst fire I've ever seen."

As Gerard sped to the hotel, cursing every traffic jam along the way, calls went out to reporters and photographers.

The *Sun*'s newly appointed managing editor, Gary Thompson, arrived at the office. Thirty-four-year-old Thompson had previously been the paper's city editor. This was his first day in his new job, and now he was organizing the coverage of what was swiftly developing into the biggest news story in Las Vegas history.

As reporters were located, they were deployed around the city: Harold Hyman, Scott Zamost, Monica Caruso, Penny Levin, Chris Woodyard, Jeff German, Phil Hevener, Mary Manning, Mary Mele, Bob Palm, Bill Becker and Laura Lyon. For some of them, the heartwrenching, often gruesome sights of the day would be a shattering experience. The same was true for the paper's photographers: Waite, Joe Buck, Jim Laurie, Ken Jones and Don Ploke, racing toward the scene of the disaster.

The newsman believed to be first on the scene was 007, who gives morning traffic reports from a van crammed with radio gear to listeners of station KDWN.

A local businessman who prefers to remain anonymous, 007 is always introduced by disc jockey Dennis King after King reads the school-lunch menu for the day.

That morning, 007 was excited about an interview he had taped with Robert Urich, who portrays Dan Tanna on television's *Vega$*. 007 had just run the interview and switched back to the station, when he heard from "Captain

Dave" Donahue, who supplies traffic information from an airplane called *Desert Eagle 1*.

"I'm just coming up from the North Las Vegas Airport," Donahue said. "It looks like there's a lot of dust coming up over by the MGM."

007 didn't think it was dust. There wasn't enough wind stirring to kick it up. He suggested that the pilot fly over that way and see.

"My God," Donahue said. He was now on the air. "This is serious. There's a lot of smoke coming from the MGM."

007 turned on his emergency flashers and leaned on the horns, with his lights flashing. He figured the police would either write him a master ticket for the year or leave him alone.

He made it to the MGM Grand in six minutes and saw Donahue flying a pattern over the hotel roof.

"Hold that command position unless Nellis sends a fighter plane to shoot you down," 007 bellowed into his microphone. He knew that local radio and television stations would soon be dispatching planes and helicopters to the scene.

Firemen were laying hoses, and smoke was boiling out of the main doors in a column twenty feet high and a hundred feet wide.

Metro police and Nevada Highway Patrol cars converged on the Strip and Flamingo Road. Firetrucks surrounded 007's van. He opened the sunroof to get a better look and saw palm trees on fire. Hot tar was sputtering down, and ashes fell into the van.

007 was asked to broadcast the need for oxygen, blankets, and food. Then, incredibly, as the news of the fire

flashed over the wires, he started getting calls on his mobile telephone from San Francisco. A friend there had remembered his mobile-phone number and given it to Bay Area broadcasters. The traffic reporter's van had become a small, self-contained command post amid the turmoil on Flamingo Road.

As the sun was coming up that morning, radio station KORK news director Andrea Boggs thought to herself how slow a news day this was going to turn out to be.

She was thinking, *Today is the day we'll find out who shot J. R., and that might be our top story.* Then came the fire dispatch on the scanner, and she called the NBC network.

From that moment on, fielding calls from around the globe, Boggs anchored continued radio coverage of the unfolding tragedy. Helping her were students from her newscasting class at Clark County Community College.

Phones were ringing nonstop at KORK. There was no time for private thoughts, but as the number of fatalities grew, she wanted to cry.

Television, magazine, and newspaper reporters from other cities began their pilgrimage to Las Vegas. Some arrived within an hour. Before the day was over, more than a hundred media representatives would be on the scene.

Elsewhere in the city, in their apartment less than a mile west of the hotel, Dan and Liz Newburn were getting ready to leave for work, when they heard a bulletin on television about a fire at the MGM Grand. Dan Newburn, a doctor of theology and the *Sun*'s religion editor, grabbed his tape recorder and camera and dashed with his wife to their car.

As they neared the hotel, he stopped to shoot a few frames. Traffic jams halted their progress a block from the MGM, and Newburn jumped out of the automobile, telling his wife to go on to her job. It was 7:25.

Running the last block, breathing hard, he was describing what he saw into his miniature tape recorder. At the fire scene, as flames leaped from the front of the MGM, the air was full of the sounds of breaking glass and the screams of people in the windows.

A firefighter called instructions through a bullhorn to the growing number of hotel guests frantically signaling from balconies, then turned to the crowd on the street and asked if anyone could speak Spanish. Several volunteers stepped forward and were pressed into service as interpreters.

In the hubbub, Newburn went up to a dazed-looking man holding his head. Taking the man's arm, the newsman asked, "Were you in there? What did you see?" Steadying himself, the distraught man said, "I was inside the casino playing 21, when I thought I smelled smoke. I picked up my chips and started towards the door. All of a sudden, the roof fell in, and this giant ball of fire started toward me. I ran, and as fast as I could run, the fireball was gaining on me. I just got out in time!"

He reached into his pocket and pulled out a handful of twenty-five-dollar MGM chips and asked, "Where do I cash these in?"

Newburn patted his shoulder and advised, "Hang on to them, they'll probably become collector's items."

Interviewing people in the crowd, he tried to talk to three cocktail waitresses huddling together on the street. They were crying and trembling. Between sobs, they explained, "Our boss told us not to talk to anyone." As more fire units and police cars converged on the scene, Newburn put his arms around two women in nightgowns and bathrobes who had just been lowered to the ground in a construction worker's spider cage.

Helping them from the cage, he walked them across the street to the Barbary Coast Hotel. As he gently guided them to chairs, they collected themselves and recounted the details of their escape.

"We were asleep on the 17th floor," they gasped. Awakened by shouts, they put on their robes and started down the hall, choking in the dense black smoke. People were running and bumping into them as they groped their way along the corridor. When they reached an exit, it wouldn't open. Their own room door had locked behind them, so they felt their way along the hall, trying every door until they found one left ajar by a fleeing guest. Inside a room once again, they flung a chair at the window. Leaning out over the jagged glass, they were screaming for help when spotted by an ironworker in a spider cage.

"Over here, please! Oh, please!" they called, and he inched the cage to their window and scooped them aboard.

Newburn left the pair in the hands of paramedics, then telephoned the *Sun* newsroom. "C'mon in," said Barrows, and the religion editor ran to the nearby Flamingo Hilton and from there caught a cab back across town to the office, where he would help anchor coverage of the story.

5 . . .

The Disaster Plan

When the story of the MGM Grand Hotel fire is re-counted in years to come, perhaps no aspect of it will be more ironic than the fact that some of the hotel's guests first heard of the fire while watching ABC's *Good Morning America* television program, which originates in New York City.

The hotel's switchboard operators were forced to leave the board between 7:15 and 7:20 A.M., as the heat became overpowering. The direct-dial phones in the guest rooms continued to function until 8:20 A.M., when the searing heat actually melted the switchboard's components. For some, the telephone was a lifeline to the outside world.

At Central Telephone-Nevada's downtown Las Vegas offices, operator Marilyn Braner got a call from a Spanish-speaking man trapped in his room on the twenty-fourth floor.

"What do I do?" he asked in a desperate tone. "The fire department tells us to stay in our rooms, but I'm choking on all the smoke."

Braner told him to put wet towels under his door and answered another frantic call.

"My God! What's going on? I can't even see the phone in front of me!" a frightened man told operator Julie Wasleski. The phone dropped. Then, silence.

"My husband jumped out the window."

"My wife is unconscious. Help me."

"What am I going to do?"

The fire department had left instructions for all the operators to tell guests to remain calm, lay a wet towel in front of the door, and wait for someone to lead them to safety.

Operators got the chills every time the monitor flashed on. Hundreds of people were depending on them for help.

Rox Ann Wysocki, a Central Telephone network monitor, got to work at 7:15 A.M. One of the women in the office had her radio on, tuned to an all-news station.

"A fire at the MGM Grand Hotel on the Strip has engulfed the bottom two floors."

Things started to snowball. The status board, which has an indicator light for every trunk line in the United States, came to life. It lit up like a Christmas tree. Local telephone traffic poured over the circuits.

In the switching room, amber alarm lights were flashing at the end of each marker frame as the tidal wave of long-distance calls rushed through downtown Las Vegas. The mechanical switching units, mounted on the frames and connected to one another by miles of color-coded wires, clicked frantically like a swarm of locusts.

Telephone traffic analyzer John Seymour, who monitored the activity in the 4-A toll switching room, recalled the last time he had seen similar telephone equipment act that crazy. That had been when he was employed with the telephone company in Colorado Springs, when President Kennedy was assassinated in Dallas in 1963.

Fighting his way through city traffic, jumping median strips in his four-wheel-drive vehicle, was Dr. John Batdorf. He had developed the Clark County civil disaster plan back in 1970.

Shortly after 7:00 A.M., he had been pulled from his morning shower and told to report to the MGM Grand. Batdorf is one of eight doctors designated to be first on the scene in the county disaster plan.

His home was near the MGM Grand, but the trip still took him twenty minutes. As he made every shortcut conceivable, his first thought was, *This is what we've been predicting for years. I hope everything works.*

Making everything work was dependent on a disaster plan covering every contingency and on the skill gained through regular disaster drills by city hospitals and emergency services.

Just two days before, there had been such a drill. The hypothetical situation had been a midair crash involving two airplanes, with flaming debris falling on a convalescent home.

The problem of dispatching and treating fifty casualties from the mock disaster had been efficiently handled.

The MGM Grand fire, however, was the real thing.

As Desert Springs, Sunrise, Valley, and Southern Nevada Memorial hospitals prepared for an unknown number of casualties, fire-department paramedics, EMTs, and doctors began arriving at the scene.

The initial plan was to transport burn victims to Southern Nevada Memorial Hospital, trauma cases to Sunrise Hospital, and cardiac-respiratory patients to Desert Springs and Valley hospitals. However, it soon became apparent that the walking wounded, as well as those requiring immediate medical attention, were suffering from smoke inhalation. Therefore, patients were sent to all four hospitals.

Before the day was over, the MGM Grand fire would involve several thousand professional and volunteer people. But at that early hour, no one had any idea as to the scope of the disaster.

At Desert Springs Hospital, on Flamingo Road, two and a half miles east of the hotel, Sheila Trexler, a registered nurse who heads the hospital's emergency room, could see the towering smoke. *We're going to have big trouble,* she thought.

When the announcement sounded, "The disaster plan is now in effect," some employees groaned, "Oh, no, not again. Not two in the same week," as they raced to their emergency posts. But this was no drill.

In the triage area set up in the lobby, patients were swiftly evaluated and dispatched to appropriate departments for treatment. In the hectic first minutes of the crisis, nurse Marilee Peraza found herself trying to persuade a Mexican guest to give up the fireman's jacket he was wearing. "No, I'm going to keep it," he insisted. "This is a souvenir of the fire."

Nursing director Davida Lewin, escorting a patient into the hospital elevator for a trip to the second and topmost floor of the building, found him shaking and visibly frightened.

He had been rescued from the hotel's eighteenth floor, and now demanded to know, "How high up are we going?"

"Only one flight," she assured him.

At 7:25 A.M., Karl Muninger, a Clark County emergency-medical-services technician, turned off his television set and headed for work.

In his car, only a few blocks from home, Muninger heard a news bulletin on a local radio station.

"We've been advised that there is a serious fire at the MGM Grand Hotel. Metro police and county fire-department officials have asked motorists to avoid the Flamingo Road–Strip intersection this morning."

Muninger was not particularly alarmed by the report. There had been a number of hotel and motel fires within the past two weeks. No one had been killed in any of them, and the most severe injury had been a broken ankle.

But as he neared the Clark County District Health Department, he could see the MGM Grand high-rise enveloped in a black shroud of smoke.

When Muninger arrived at his headquarters, officials had already activated the Med-Alert Emergency Medical Services (EMS) disaster response. A deputy health officer, Evelyn McColl, set up a communications command post at the health department.

Muninger and McColl then headed to the scene.

Mercy Ambulance general manager Bob Forbuss had designated Flamingo Road, near the hotel's Grand Portal, as the main triage area. Forbuss was acting as triage officer when Muninger and McColl arrived at the besieged Strip resort.

Although the disaster plan called for a single medical command post, set up on the north side of the hotel, the prospect of five thousand or more guests and employees requiring medical attention necessitated three separate areas, north, south, and southeast of the MGM Grand.

Muninger's first task was to clear away spectators. He was assisted by Metro police officers. Next, a temporary receiving area for the walking wounded had to be found. It was set up in the Keno lounge at the Barbary Coast, where hundreds of terror-stricken evacuees had already gathered.

Fifteen of the city's seventeen paramedic units were rushed to the MGM Grand. Two were left to cover the rest of Las Vegas. All ambulances within seventy miles were called in. Drivers were instructed to check in at the north command post.

Forbuss contacted the Clark County School District and arranged for fifty schoolbuses to be routed to the fire scene. They would be needed to transport hotel guests to the Las Vegas Convention Center, as set forth in the master emergency plan.

Four disaster physicians were at the scene by 8:30. Two of them were directed to supervise activities at the north triage area. Emergency-medical-services physician Ralph D'Amore and Las Vegas Fire Department paramedic liaison officer Rex Shelburne ran the operation at the south triage area.

Clark County Emergency Management Coordinator Ken Ryckman and emergency-medical-services field representative LaRue Scull, who served as triage officers at the southeast post, radioed the north command post: "We've got mostly noncritical patients here."

Otto Ravenholt, Clark County's chief health officer and county coroner, arrived at the north triage area at 8:30. He took charge of the prehospital medical effort.

The East Hall of the convention center was designated the primary receiving area for hotel guests not in need of hospital care. The first buses of patients were loaded and off at 8:45.

Requests for supplies poured into the command post.

"We need additional oxygen tanks, regulators, needles, syringes, tubing, IV solutions, medicine cups, blankets, insulin, and other medication," Muninger told staffers at the Health Department command post. Ten physicians were then at the scene.

Clark County Fire Department paramedic liaison officer Mike Verrilli and American Ambulance president Ryan Johnson arranged the ambulance and bus flow in such a way as to reduce traffic congestion on Flamingo Road and adjacent streets.

Supply items were delivered within ten minutes after they were requested. In addition to the disaster frequency, Med 9, Med 10, Fire Frequency 4 and the Mercy Ambulance frequency were used at the scene.

* * *

At 9:05, the fatality count stood at eighteen. Additional body bags were requested from the coroner's office.

By 10:45, three hundred and fifty victims had been transported to hospitals. Blood-gas tests on incoming patients showed high levels of carbon monoxide. These findings were radioed back to the physicians at the scene.

Ambulance traffic had slowed up somewhat, but the tragic statistics were just starting to come in. Thirty-seven dead . . . sixty-four dead. . . . Rumors circulated that more than a hundred would be found dead inside the steel-and-glass tower.

To Muninger, it seemed evident that the remaining victims would be transported from the scene in a coroner's car.

Betty Bres of the Red Cross heard the news as her husband turned on his car radio while backing out of the driveway at 7:30 A.M.

Hurrying into the house, she turned on the radio and listened while she changed clothes. The fire was a bad one. Already, there were reports coming in of people hanging from windows, screaming for help.

She called other Red Cross workers, then collected first-aid equipment and got in her car. She knew where to go.

Harry Christopher, Metro police helicopter pilot, placed a call on the police radio asking that Flight-for-Life—a medical evacuation helicopter based at Valley Hospital and used to transport patients from remote areas of southern Nevada and contiguous states—come to the MGM Grand

The upper floors are engulfed in smoke.

Smoke bellows out from the roof of the MGM. Guests can be seen at windows and on balconies.

The fire as seen from the hotel swimming pool.

Guests gaze up at the rescue helicopter. Photograph taken from the swimming pool.

Some of the guests hoped to escape by using knotted-together sheets as ropes.

Rescue operation.

Hotel guests climbing down fire-ladder to the street and safety.

Many of the MGM guests smashed windows. Note lower left, where a table, tied to a window post with a sheet, was used to break glass panels.

Guests on the balconies being rescued via the spider cage.

"Jolly Green Giant" helicopter rescuing a guest from the roof.

A guest swinging from a military helicopter.

Smoke inhalation victims being given oxygen

A fire victim is carried from the scene by CCFO paramedics and construction workers

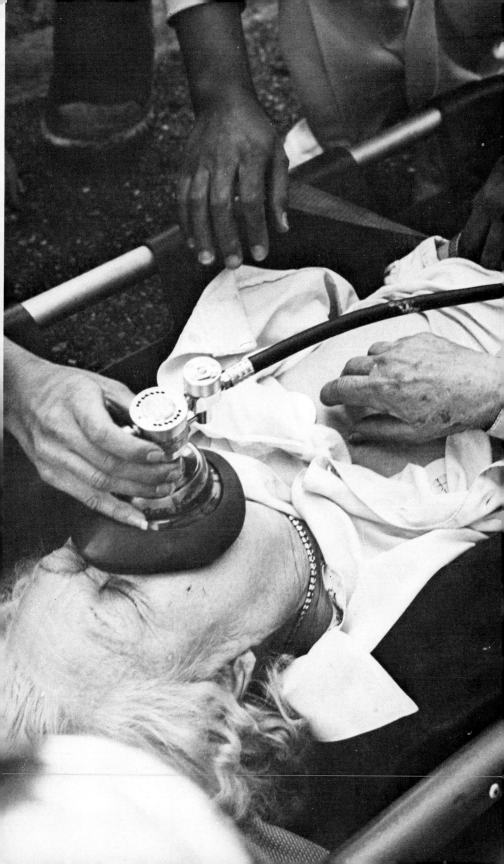

Another guest being helped to the doctors' station.

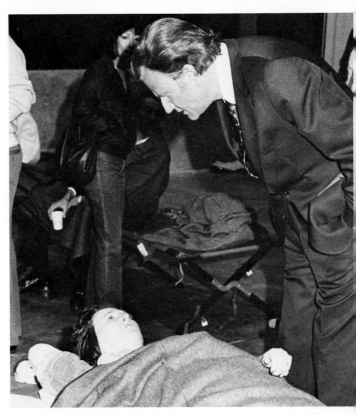

The Rev. Billy Graham consoles a fire victim at the Las Vegas Convention Center, where an emergency medical station had been set up.

to airlift people from the roof. He also called for all available civilian and military helicopters.

Gary Skelton, a Metro police officer who works out of the department's substation at McCarran International Airport, heard the broadcast and called Ray Poss of Silver State Helicopters at home.

Poss lives six miles from North Las Vegas Airport, his headquarters, and was at the airport and in his Bell Jetranger within twelve minutes. He could see the black column of smoke and knew that every minute counted.

When Poss got to the MGM Grand, Flight-for-Life pilot Paul Kinsey and Metro Police's Air 1 were in the area. Harry Christopher had made the first twenty trips to the roof, where hundreds of hotel guests, choking in the encroaching smoke, were rushing the helicopters.

The rule was women and children first. A police officer and his dog made sure that the rule was enforced.

The smoke was intense, and it was hard for the chopper pilots to see anything within several hundred feet of the MGM Grand's roof. The pilots arranged to have one helicopter hover near the roof to blow the smoke away while the others attempted to land.

In the rooms, people waved as the helicopters went by, trying to get the pilots' attention. Others signaled thanks for a few breaths of fresh air as the rotors beat away the acrid smoke.

Trapped in guest rooms, desperate men and women held up makeshift signs: MY HUSBAND HAD A HEART ATTACK. . . . MY WIFE IS PREGNANT.

The civilian helicopters had no equipment for balcony rescues. That would have to wait until the air force arrived.

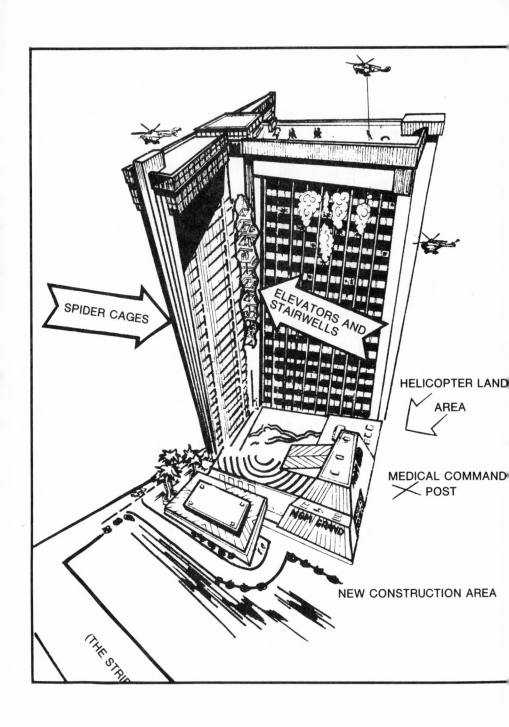

SPIDER CAGES

ELEVATORS AND STAIRWELLS

HELICOPTER LAND AREA

MEDICAL COMMAND POST

NEW CONSTRUCTION AREA

(THE STRIP)

MGM GRAND

Artist's rendering of the MGM Grand, showing entrances and location of the casino

Cutaway, showing where activities took place

Smoke was pushing up from all sides. Many of the guests feared that the roof would collapse any moment.

Poss's helicopter had room for four passengers, one in the front and three in the back.

Many people were more concerned about their luggage than about their own safety. Some threw their suitcases and personal belongings into the chopper before they climbed on board. It got to the point where Poss and the other helicopter pilots had to throw luggage back out on the roof to make room for people.

Although Poss had been flying choppers for eighteen years, he had never seen anything like this. It was awesome. He saw people collapse on their balconies, and he knew that they were either dead or dying.

For some MGM Grand Hotel guests on the lower floors, escape or rescue came early. These lucky people, in various stages of dress, stumbled into nearby hotels.

At the Maxim Hotel and the Barbary Coast, both on the other side of Flamingo Road, barefoot, half-dressed men and women started filing in before 7:30, to the amazement of employees and guests.

Sharon Coakley, just finishing her midnight-to-eight shift in the Maxim change booth, had heard the sirens outside. But because Fire Station 11 was just down the street, sirens were not remarkable.

A security guard had mentioned something about a fire at the MGM, but no one at the Maxim imagined that it might be something major.

Then people dressed in nightclothes started entering the

Maxim and Barbary Coast lobbies, demanding telephones. One dazed woman sat down in the Maxim keno lounge and stared into empty space. Still, no one at the three-year-old resort fully realized what was happening across the street until a security guard looked outside.

At the Barbary Coast, opposite the MGM Grand's blazing front there was a clear view of the flames. Owner Michael Gaughan shut down his gaming operation as personnel pushed tables aside to make room for a first-aid station.

From across the Strip, at Caesars Palace, hotel employees rallied to help. The Dunes Hotel set up thirty-one beds in its convention hall and sent food and coffee to evacuees and firefighters.

Terry Wall rushed down from the nearby Flamingo Hilton Hotel and helped mobilize the removal of cars from in front of the MGM Grand.

Salvation Army Major Rex McCulley lives near the MGM Grand. Hearing a bulletin about the fire on television, he looked out the window and saw the smoke. Salvation Army workers were soon at the scene with a coffee canteen on Flamingo Road. McCulley then sped to the warehouse for emergency stockpiles of clothing. They would be needed.

Roused by a phone call early that morning was Claudette Shawbridge, NET Council project director, whose organization, with headquarters in the Las Vegas Metropolitan Police Department's main station downtown, coordinates the volunteer resources of the three local Radio Emergency

Associated Citizens Teams, Southern Nevada Volunteer First Aid and Rescue, and the Nevada Community Assistance Patrol.

This organization is ranked one of the best of its kind in the United States.

Her caller was Jim Murphy, in the final hour of his graveyard shift monitoring of CB emergency Channel 9 for NET.

"Jim said there is a fire at the MGM," Shawbridge said later. "I was still half-asleep and didn't think much of it. I told him to call Gene Alexander, chairman of the NET Council, and hung up the phone."

But then: "Right after I put the receiver down, I jumped up in my bed. A fire at the MGM! It might be serious."

She reached for the phone and dialed the NET Control number. Busy. She punched the seven digits again. Still busy.

The office was probably being deluged with calls, she thought, so Shawbridge phoned the Metro police watch commander on duty, Lieutenant Harold Rowe, and asked him to relay a message to her people at NET Control.

"Tell them to stay put. I'm on my way over."

Meanwhile, several miles across town, John Jones and his wife, Renée, had just arisen from a good night's rest.

John is a day-shift dispatcher for Valley Hospital's innovative Flight-for-Life program. He is a certified emergency medical technician (EMT). Renée is a licensed nurse.

When the couple is not at work, they volunteer by operating one of Southern Nevada Volunteer First Aid and Rescue's (SNVFR's) three ambulances at Craig Road

Speedway, an automobile race track in the desert northwest of Las Vegas.

"I called the office to talk to Jim Lewis, the dispatch supervisor," John said. "He told me that the MGM was on fire and that the helicopter was there evacuating people."

Claudette Shawbridge ran into the NET Control room at 8:05.

"When I got there, we had just gotten a call from John Jones requesting that all available EMTs respond to the MGM," Claudette said. "Then we received a call from Don Cruz with the Red Cross, who asked all available NET personnel to report to the Convention Center."

Among physicians reporting to the Medical Command post at the hotel were Dr. Elias Ghanem, Dr. Lonnie Hammargren, and Dr. William Berliner, who were treating cardiac arrests and the many cases of shock, high blood pressure, and smoke inhalation.

Hammargren, a neurosurgeon, described many evacuated people as "the walking wounded," as they sagged in the arms of rescuers.

For Berliner, as he labored over fire victims, the disaster had a poignant extra dimension: His daughter had been at work in the MGM Grand when the fire erupted, and he had no way of knowing what had happened to her.

Police cordoned off the streets so emergency vehicles could get through. Ambulances delivered two hundred fire victims to Sunrise Hospital, Nevada's largest medical center, on Maryland Parkway. There, medical staff and hospital volunteers speedily expedited treatment.

Paperwork would be at a minimum that day—names

and addresses were taken, and the hospitals were instructed by MGM Grand officials to have bills sent to their local insurance adjusters.

In a reception area, television reporters captured the taut scene as families waited for news.

A memorable moment was captured when TV cameras focused on a Mexican-American guest who had slid down a rope from the eighteenth floor. Miraculously unhurt, except for minor rope burns, he beamed at his good luck.

6 . . .

Heroes of the Day

One of the first to escape the building was Mark Macud-zinski from San Francisco.

"I was sleeping and heard some commotion; some voices down the hall," he told the *Sun*.

"I smelled something a little funny, and I walked to the door and looked down the hall. There was a black wall of smoke coming down the hallway. I ran the other way toward the stairwell. I didn't even have a pair of pants. A guy from Tennessee gave me his pants."

Another early escapee was Len Brown, in Las Vegas for the computer convention. He was on the phone to his children in Houston, when the telephone went dead.

What a hotel, he said to himself. *Even the phones don't work.* He put the receiver down.

He heard voices outside and wondered what was going on. People were yelling, "Don't leave your room!" and "Don't panic!"

With radio his only link with the outside world, Brown listened for reports. He heard windows breaking. As thoughts of the movie *The Towering Inferno* flitted through his head, firefighters arrived to lead him out.

Mr. and Mrs. Melvin Dickey, from Seattle, had retreated back into their sixteenth-floor room after groping through about fifteen feet of the hallway.

They wetted down several towels and tucked them under the door. Then a firefighter with a bell, a flashlight, and an air tank appeared at the door and led them to safety.

On the seventeenth floor of the MGM, Chris Gripp and Steve Toburen, owners of a computer business in Durango, Colorado, crowded into a room with about thirty other people. It was the only place on the seventeenth floor not filled with lethal smoke.

Among those in the room were two little brothers, one twelve years old, the other four months old. Seeing the youngsters made Gripp and Toburen think of their own children safe at home. They had at one time considered bringing their families with them to Las Vegas. Now they were glad they had not.

Then as the sirens screamed and glass from broken windows crashed into the street below, a soot-blackened man in the small construction rig called a spider cage came into view outside the window. It was ironworker Chris November, one of the construction crew from the hotel's new addition.

In the crowded guest room, Toburen strapped the baby into the parka of the twelve-year-old, as the boy's parents gave him money to buy formula. Reaching out a muscular arm, November took his small passengers aboard the swinging cage. With a couch cushion covering the children's heads to protect them from falling glass, November began his descent.

"It doesn't matter what happens now," said the father, as he watched his children disappear down the side of the building.

There were heroes that day. As the deadly fire swept everything before it, paramedics, firemen, police, helicopter pilots, MGM Grand employees, and passersby summoned from within themselves incredible strength and courage as they repeatedly risked their own lives to save the building's thousands of occupants.

And none did more than the ironworkers, about seventy men of Ironworkers Local 433 who were working on the new highrise addition. By seven o'clock Friday morning, most had arrived at the resort's new construction site, where a skeleton of steel girders stood half-finished adjacent to the towering hotel. Some were preparing to mount scaffolds leading up to the twenty-sixth floor.

Suddenly, smoke was billowing forth from the front casino entrance, and the ironworkers began yelling, "Fire!"

"The people in the hotel rooms looked out their balcony windows to see what was going on," said Patrick Gallegos, a member of the John F. Beasley ironworkers' crew. "One man and his wife walked out on their balcony about the eighteenth floor. They stood there looking down at the smoke but weren't moved. The man waved at us and lifted the glass he had in his hand as if to cheer us and just stood there for a while."

Some of the men on the derrick floor cursed at him.

Another man and woman ran out on their balconies stark naked. They saw the smoke and quickly disappeared into their rooms.

Five minutes passed. "Then we saw a ball of fire break out between the front entrance canopy and the hotel entrance door. Hell had broken loose."

While screams from frantic guests filled the air, some ironworkers quickly helped firefighters with hoses.

Others, like Chris November, Jerry Hoskin, Hubert Edwards and Rusty Moore, went up and down the T-shaped tower on scaffolds, plucking guests from their rooms.

Many ironworkers, with wet towels strapped around their necks, ran inside, through the smoke-filled hallways, kicking down doors and leading guests to safety down stairways.

They stumbled over dead bodies, fought off poisonous gases, carried dying victims over their shoulders, and watched helplessly as some died in their arms.

Christopher November, known as C. J. to his friends, was among those who "ran all over the place helping people" before the first major wave of firetrucks arrived at the scene.

"The firemen tried to get people down as best they could, but, they could only reach the eighth floor," said C. J. "I thought to myself, *This isn't enough*. People were leaning over their balconies, screaming in a panic for their lives. I knew something had to be done."

November jumped into the cramped spider cage, turned on the power, and took it up the side of the building.

"I went up one time and was told not to go again." The

steel cable might overheat, he was warned. The motor might burn out.

"But when I looked up at the building and saw the faces of those people, scared for their lives—you know someone has to do something."

November was getting two, three, sometimes four people down on each trip.

"Everybody wanted to be saved. Everybody wanted to come down," he said. "There was one man who didn't wait for me. He tried holding on to some sheets he had tied together, but his body weight was too much and he fell all the way—twenty stories. A young lady tried the same thing. There was no way I could stop her. I wish I could have."

For a while, all C. J. could hear was frantic calls for help. "There were people screaming and just panicking from one end to the other. I had to tell people what to do and what not to do. I did the best I could to tell them they'd be all right."

Tables and chairs came crashing through windows. Shards of glass rained down on C. J.

"I was going up to get another couple, when the power went out. Now I was stranded and had no place to go, but to try and climb up the line and get on one of the balconies." A good number of the more than fifty people C. J. had taken to safety watched with fear as he pulled himself along the cable.

He made it.

Jerry Doyle remembered "carrying a woman several floors, thinking, *I am not going to make it another step.*

But God somehow gives you that extra strength you never thought possible."

Patrick Gallegos told of rescue efforts on the upper levels of the hotel.

"We got to the sixteenth floor, and no one had been there yet. So we felt the door for heat. It was cool, so we broke the door in, and smoke poured out of the hallway. . . . We could hear people coughing and yelling for help. We knocked on doors and showed people to stairwells. Some were in shock. Some were dead.

"From the eighteenth and nineteenth floors and up, it got worse. We could barely breathe. We had to break into nearly every room. There were a lot of elderly people on those floors who were having trouble breathing. Most of those people were suffering from smoke inhalation. They were weak and scared.

"My arms and legs were hurting like hell from carrying people, but I couldn't stop."

It was a tragic scene, too, for Pat's brother Ben Gallegos.

"I walked back to the center of the building and met one of my co-workers. After walking a little ways, we found a woman laying on the floor by the elevator. We picked her up and put her on his back, and he carried her down the staircase. Later he told me she had died.

"Breathing smoky air through a wet towel and trying to keep the poisonous smoke from doing too much damage, I got to the roof, where a helicopter crew gave me a ride down to the ground.

"On the way down, I was grateful for my life, and I thought of those I had helped. I was sad for those who didn't make it. I thought of my family, my children. . . . I had helped children, young people, old people, and I felt good. We all had something to be grateful for this Thanksgiving all our lives."

Ironworker John Scott described the fire scene, with its mushroom cloud of heavy black smoke hovering above, as the "darkest day" he had ever seen.

"There were so many people crying for help. And all we could do was tell them to follow the stairs."

Ashley Moore, another ironworker, watched in horror as one woman, attempting to climb down a rope, fell to her death from the fourteenth floor.

William Armstrong was another who helped carry people to safety. "I carried people out, until my arms were too sore to carry any more. As we neared the first floor, where we could see a little light and breathe easier, people started screaming for joy."

Fred Kiely told of death on the twenty-fourth floor.

He went down the hallway with his buddies, kicking in doors. They found two people near death in one room. One of them died before they could get him to safety. Before he died, he asked if his wife was all right. Kiely looked and saw that she was dead.

"All day long, everywhere I went, there was a beehive of ironworkers helping people," says Jerry Hoskin. "They instinctively did not panic and carried out their acts of heroism without personal regard for their own safety. It's damn nice to know each and every one of these people personally."

More than half of the ironworkers had to be treated for smoke inhalation after their daring rescues that morning. Hubert Edwards spent four days in the hospital.

* * *

Choreographers Winston Hemsley and Rich Rizzo had gone on to join the impromptu parties springing up among the cast of *Jubilee* after the late rehearsal. Then, at 6:00 A.M., they sat down for breakfast in the hotel's coffee shop, before going up to their room on the twentieth floor.

Sleep overcome them fast once they hit their beds. But less than an hour later, Winston awoke with a dry throat. It was uncomfortably hot in the room. Eyes still shut, he thought about getting up to turn down the thermostat.

First blinking, then fully awake, he opened his eyes, to see a cloud of smoke enveloping the room. He jumped out of bed and dashed to the door. The corridor was black with smoke. There were no people, no alarms ringing. He ran for the bathroom and grabbed all the towels, stuffing them under the door and into the ventilator openings. Then he woke up Rich.

"Don't panic," he warned, "but I think the hotel is on fire."

"A fire!"

In spite of Winston's urgent calm, Rich did panic. He flew out of bed and smashed a chair through the window. Together they broke the rest of the glass, then pulled down the draperies, drenched them, and stuffed them under the door. Next, the bathtub. They filled the tub and bathroom sink with water and left the taps running, so the overflow gradually soaked through to the bedroom.

Winston is a thirtyish black New Yorker, and while growing up, he had seen the worst that city has to offer. He had also seen the best and tapped his way to acclaim in Leslie Uggams's starring Broadway show *Hallelujah, Baby*.

From there, he came to Las Vegas to a featured spot in the MGM Grand's *Hallelujah Hollywood* with partner Alan Weeks. And all his life, whatever he was doing, he was a movie fan. He once said, "When I first came out to Las Vegas, I already knew what the desert would look like. I'd seen it all in the movies, many times."

So at 7:30 on the morning of November 21, groggy from lack of sleep, he was thinking about a movie, and it was *The Towering Inferno.*

He reran the film in his mind. What had those people done to survive? "The most important step," he told his friend, "is to stay calm and don't move from where you are until you appraise the situation." But he had to convince Rich of that.

Rizzo was a total wreck. He ran for the hall, but Winston tugged him back and propelled him to the window. He then forced Rich to look up, at a dead man hanging over the sill above. He forced him to look down, where a dead woman lay sprawled on the ground.

"They're dead, dead, dead!" he kept repeating, to bully Rich into collecting himself. They talked about the show.

Their window overlooked the roof of the Ziegfeld Theatre, scene of the triumphant rehearsal only a few hours earlier. "Well, there goes the show," remarked Winston. "All those months of work, going up in smoke." They were thinking about Donn Arden, the producer of *Jubilee,* and what this would mean to him. They were also worrying about Arden and Miss Bluebell, a former showgirl who was Arden's co-producer. Winston and Rich were youthful and fit, and they were having trouble breathing. How was this affecting older people?

When they saw helicopters landing on an empty lot below, he told Rich to watch them. If the choppers started taking off, they'd know they were in big trouble.

"How do you know?"

"Because the same thing happened in *The Towering Inferno,*" Winston replied.

Moments later, as the helicopters rose into the air, Winston groaned, "Oh-oh—this is it. We'd better start figuring out an escape." Looking down, they saw the smoke coming up past the first two floors of the building. They decided that when it reached the sixth floor, they'd put wet towels over their faces and try to make it up to the roof.

Around them, shouts and screams came from neighboring rooms. Then a curious sense of detachment took over, and as they watched and waited, they talked about how strange it was that it was men, more than women, who were panicking and crying.

Forty-four-year-old Richard Amador, a member of the Alhambra, California, Board of Education, and his wife, Theresa, forty-two, the city clerk of Monterey Park, California, were in Las Vegas for a convention on school budgeting. Their group had planned to stay at the Las Vegas Hilton, but because the Hilton was overbooked, they were transferred to the MGM Grand.

Thursday night, the Amadors left a wake-up call for 7:00 A.M. "Thank God for that call," Richard said later. "I was in the bathroom shaving, when Theresa called from the other room that she smelled smoke." Looking out the window of their sixth-floor room, they saw a cloud of smoke that had formed suddenly.

The door wasn't hot, so they wrapped wet towels around their heads and were preparing to leave their room, when suddenly the door flew open and two people burst in from the hall. The strangers picked up an armchair and threw it at the window. With smoke from outside pouring into the room, the Amadors dashed into the corridor. It was a pitch-black hell, filled with desperate, screaming people.

They dropped to their hands and knees and started crawling through the blackness.

"Stay down!" Richard yelled to his wife and the people around them. "Stay close to the floor," he kept warning as, hacking and coughing, they looked for an exit.

As the Amadors crawled on their hands and knees, screams filled the air.

"Stay down close to the floor!" Richard yelled.

They got to an exit, and it was locked. They found two more exits, and they were locked.

Somehow, they kept on crawling, trying to escape smoke. Gasping and coughing, they finally found a stairwell and stumbled down. But when they thought they were free, they found themselves trapped in a smoke-filled patio. A mob of people was pushing its way into the small space, but there was only one way out—to go over the wall. With one last mighty effort, they made their escape.

As the hours ticked by, Winston Hemsley and Rich Rizzo were entertaining themselves in their room on the twentieth floor by imagining how they would choreograph a disaster. It was Silly Time—it was a time to make jokes or give up.

The police helicopter flying past, with its loudspeaker blaring, "Stay in your rooms—the fire is under control," brought them a measure of reassurance. Laughing, talking, dancing out movements of their disaster choreography, they had no sense of time. But looking out the window, Rich noticed that the smoke was dissipating and that people were no longer gathered on the lower balconies.

They must be getting them out, Winston thought. He cautiously opened the room door and peered into the smoky hall. Visibility was slight, but he could see bodies sprawled on the floor. He quickly closed the door and told Rich things were looking better. A moment later, a firefighter, almost completely overcome by smoke, staggered into the room.

The two men, both perfectly fine, helped him over to the window for some gulps of fresh air, then wiped off his blackened face.

"We'll help you find some other people," they offered, but the firefighter, catching his breath, shook his head and led them out into the hall. There they were joined by a couple from a nearby room, and they watched, unbelieving, as the man grabbed the firefighter's arm and took off down the corridor, leaving his wife behind on her own. Winston took the woman's arm and helped her along.

Then came the worst part of their ordeal—stepping over dead bodies on the floor. Rich had never seen death before, and he thought, *I never knew a dead body could look so small.* Then he started crying as he realized that the body he was looking at was that of a young girl.

The firefighter led them down a staircase, then through a hole knocked into an outer wall and on out to a lower roof where ladders waited.

On the ground, many of the stage crew and cast were helping with fire victims, and they gave a mighty cheer as they saw the two choreographers emerging on the rooftop. At once Rich and Winston, show business to the core, did a full stage entrance, grinning and waving as they clambered down the ladder. But once they were safe on the ground and embraced by their friends, the euphoria vanished. Shock set in, and Rich was shaking and crying.

Winston felt himself getting sick, but this was no time to collapse. "C'mon," he said to Rich. "We're okay," he told the medical aides who flocked around to give them oxygen. Rich, composed again, ran with Winson through the crowds to the pool area. At the hotel's southeast entrance leading to the Arcade, they sneaked past a cordon of security guards and went back into the hotel with the idea of finding Donn Arden and Miss Bluebell.

"It was a stupid thing to do," Winston would say later, "but we were emotional and not thinking too clearly." Their progress was halted by another guard, who ordered them back outside. Asking around, they learned that Miss Bluebell and Arden were both safe, and so was choreographer Tom Hansen. He was rescued early in the day and had already telephoned news of his safety to friends in Los Angeles.

Only later did they discover with horror that the show team did have a casualty. Fifty-three-year-old Teresa Levitt, assistant to designer Bob Mackie, had been staying in the hotel on the twenty-second floor. She was known as Terry to members of the cast. Her body was found in the twenty-second-floor foyer, face up, her head near the elevator door.

7 . . .

The Escape Continues

At 7:10 A.M., on the hotel's sixteenth floor, Marv and Carol Schatzman were up and dressed. They had come to spend a few days in Las Vegas after a trip to California to see their son play in the University of Santa Clara football season's final game. They fidgeted as they waited for a bellboy, because they had to catch an 8:15 A.M. plane home to St. Louis.

Because they were running late and no bellboy had arrived, Schatzman decided to go down to the lobby and check out. As he strolled to the elevator, a waiter ran past and asked to use his room phone.

"Sure, come on in," said Schatzman. He was just beginning to see some smoke drifting along the hall. Nervously, the Schatzmans watched the waiter trying to call the front desk. The man dialed again and again, tapped the receiver, then slammed it down. "There's no answer," he said.

"There's a fire somewhere in the building. We've got to get out of here!"

Schatzman picked up their luggage, and the three ran down the broad corridor toward a fire exit. When they flung open the door to the stairwell, smoke poured out at them.

"This way! There's another exit!" the waiter yelled, and they shut that door and retraced their steps. By then, the hallway was darker than dark. It was pitch black.

As they felt their way along the corridor, Carol was beating on room doors with the handle of her umbrella. "Fire!" she yelled. A funny thought occurred to her: *Oh-oh—if this isn't a fire, it's a felony to yell, "Fire!" in a crowded building.*

At the far end of the sixteenth-floor hallway, they found a stairwell that was clear of smoke. They started down, but as they got to the ninth floor, they met frantic people running up the stairs.

"You can't get out that way," the people yelled. "Go back up. Go up!" The Schatzmans and the waiter started up again, and after two flights, Marv dumped the luggage in a stairwell. This was no time to be struggling with suitcases.

Exhausted, gasping for breath, they kept moving upward. In the smoky stairwell, they saw people collapsing on the stairs. No one was screaming then; they hadn't breath enough to scream. Fatigue, disorientation, and the lassitude that results from carbon-monoxide poisoning had caused them to just sit down and rest. The Schatzmans and the waiter forced themselves on, stepping over outflung legs.

They had no strength to help others; they barely had the strength to keep going themselves.

When they got to the twenty-fourth floor, two stories from the top, a bearded, bare-chested man at the head of the stairs shouted down, "The door to the roof is open! Keep coming!"

With an anguished look at still more people slumped against the wall, the Schatzmans and the waiter managed to pull themselves up the last flight of steps. Then they saw the sky through the open door.

"My God!" breathed Carol, as they emerged on the roof. "Oh, my God, this is heaven!"

In the first forty-five minutes of the helicopter operation, 235 men, women and children were removed from the roof.

Then the giant CH-3s, the Jolly Green Giants of Vietnam fame, arrived. All these nine units, dispatched from Nellis Air Force Base, would not have been available except for Operation Red Flag, the war games in progress.

With their special rescue equipment, they were ready to pluck stranded guests from windows and balconies.

Master Sergeant James W. Connett, an air-force reservist from Phoenix, was one of the rescue crew lowered by cable to balconies.

Hanging in the air, hitched to a cargo strap, he would throw the other end of the cable to a terrified guest. In the sling called a boatswain's chair, people were winched up into the cabin of the hovering CH-3s.

Also rescuing terrified guests was Technical Sergeant Daniel R. Jaramillo, flight engineer aboard a Sikorsky CH-3E Sea King.

After that dramatic morning, he recounted, "Once, I was barely onto a balcony, when this pregnant lady jumped right on and grabbed me around the neck. She almost choked me to death. She wasn't scared—just glad to see me."

Not everyone was willing to risk it.

Connett said later, "Several people, especially the older women, wouldn't budge from where they were sitting. One said she was scared of heights—we were operating from two hundred feet up and higher—but I just sweet-talked them in. I told them it was the only way out."

The helicopters hovering above the MGM Grand recalled scenes from the fall of Saigon, and many of the pilots, both civilian and military, were veterans of battlefield rescues in Vietnam. They had done this kind of close work before, and their coordination was spectacular.

Said Gene Oats, a pilot with Bauer Helicopters who flew one of six civilian craft on the scene, "The FAA gave us a special frequency. Metro was in charge of operations. You don't want helicopters running into each other under these circumstances."

Vietnam had been his training ground.

Phoenix air-force reservist Captain David T. Ellis, who assisted in hoisting people up into the big CH-3s, remembers mixed reactions as evacuees were pulled into the aircraft. "On one trip, the first person in went straight to a corner and began crying. When the second person arrived, it was like a family reunion, lots of hugging and kissing. We even got a few hugs."

One woman was so happy to be alive and out of the smoke-filled hotel that she tried to give members of the copter crew wings she had pinned on her dress. She said she just wanted to give them something.

Still another chapter of helicopter history was written on that day as Dennis Mack of Action Helicopters and Gene Oats of Bauer made repeated runs to Glendale and Overton, Nevada, for additional oxygen tanks. That's a 110-mile round trip.

Carol and Marv Schatzman and the hotel waiter stood together on the broad roof of the MGM Grand, exulting in the sight of blue sky. The wind from whirling chopper blades buffeted the waiting crowd, but never had wind been so wonderful or so welcome. Then it was their turn to be flown off, and they were directed to a small orange-and-white helicopter.

Ducking low under the spinning rotors, they gratefully climbed into the craft. It had no seats in the back. A crowbar and a hardhat lay on the seat next to the grinning pilot. The machine lifted into the air, then flew sideways.

In thirty seconds they were on the ground.

On the twenty-third floor, Barry Richards, sixty-one, always an early riser, was up by seven o'clock. The California real estate developer and his wife, Trudy, sixty, had driven to Las Vegas from their San Juan Capistrano home for a brief vacation and to inspect some Las Vegas property they own. Trudy Richards, an artist, had brought her painting kit along. They planned to return home that day.

Barry was dressing, when he heard loud voices in the hall. At first, thinking some late celebrants were just getting in, he ignored the noise. When the shouts and yells got louder, he was afraid they would awaken his wife, so he opened the door to complain—and saw terrified people running along the smoke-filled corridor.

Slamming the door, he shook his wife. "Honey, get up! We've got a problem!"

The couple heard sirens, and looking down at the street, they saw the hubbub below.

Stripping the bed and wetting down every sheet and towel, they fought the encroaching smoke.

They lined the door, but smoke came through the cracks. Newspapers! Richards drenched papers and, with his wife's palette knife, forced shreds of wet newsprint into the cracks of the doorway. Smoke poured through the ventilators. He removed the grills and stuffed pillows into the openings.

Still the fumes came in, and with a heavy chair, he smashed out the window.

Their room had no balcony, but the room next to them had one. For a time, he considered trying to break through that wall. Chances of rescue might be better from a balcony than through a broken window. But how to break the wall?

He examined every possibility of escape. Wrenching down the draperies, he thought of making a rope. "No— they're only about thirty feet long," he muttered. "That won't work."

How close was rescue? They didn't know. Richards, a survivor in his lifetime of two near drownings, a plane crash in the service, and a motorcycle accident, had only one thought—"I have to save my wife." Again he reviewed the measures at hand. But there was nothing to do but wait for rescue.

Minutes passed like hours; yet hours flew by like minutes.

When firefighters burst into their room on the twenty-third floor, the Richardses had been nearly four hours wondering when help would come.

As they hurried down the stairs, they passed several still figures slumped in the stairwell.

Trudy Richards fell on the steps. When they reached the triage area, doctors ordered them taken to Sunrise Hospital, since they had both inhaled volumes of smoke.

Only then did Barry Richards find out that he had a broken toe, fractured when he had pulled down the draperies early that morning and the heavy curtain rod had crashed down on his foot.

Singing star Mac Davis spent the night at a private home maintained for MGM Grand stars, but comedian Lonnie Shorr was staying on the sixteenth floor of the hotel. He had awakened to a tumult in the hallway. Now, only a few hours after finishing the midnight show, he was on his balcony, trying to figure out what to do.

After lining his door with wet towels and yelling encouragement to guests on adjacent balconies, he began to assess his chances of getting out alive. Strange thoughts crossed his mind, such as "swinging across balconies like Errol Flynn." He prayed. Even with wet towels over his face, he was finding it hard to breathe. Then he thought, *This can't last forever.* It didn't. Rescue came when a firefighter appeared in the doorway and yelled, "Get out of here!"

Shorr grabbed a credit card and plane tickets and rushed through the smoky hall to the fire stairs. As he ran, a firefighter hailed him and said, "Take care of this lady." It was Sophie Stromberg of New York. With her husband, Dr. Samuel Stromberg, she was finding it hard to walk.

Shorr put his arm around her and kept saying, "Sophie, you'll be all right," as, step by step, the trio made their way down sixteen floors to the safety of the ground.

Stewart and Grace Krakover were from Culver City, California, where Stewart was in the computer-hardware business. Stewart had been on an arduous business trip, which was winding up in Las Vegas for the Comdex Exposition. He had arrived on Wednesday, and Grace had joined him at the MGM Grand on Thursday. The night before, Stewart had been lucky at 21.

Now they were trapped on the twenty-fourth floor. They did their best to seal all the openings, but smoke continued to come into their room. Krakover's mind was racing, trying to figure a way out of their plight. He thought, *The window!* but then stopped. He was barefoot, and he couldn't think without his shoes on. Once shod, he felt better.

Grace rummaged through their luggage, looking for her bathrobe. When she found it, she put it on. Now they were both fine, ready to start planning what to do.

Stewart's thoughts returned to the window. "Watch this," he called to Grace. He seized a big table lamp to throw through the sliding glass door, which was locked. But the lamp was attached to the large, heavy table it rested on, and he stumbled back. "Okay, that ends that," he said. Then he grabbed a clock radio. It proved to be bolted to a night stand.

The chair! He hefted an upholstered wooden chair and

began banging it against the glass. The chair fell apart, but the glass remained intact.

Stewart yelled to his wife, "This is just like *The Towering Inferno*! I'll try another window." This one shattered open. By then the smoke was so thick that they couldn't see the bed.

For the next two and a half hours they sat on the floor amid the broken glass, with their heads out the window, watching the activity outside. They were prepared to spend the whole day in the little world he had created by breaking the hole in the window.

After their second night in Las Vegas, Bill and Dona Bonham and Tom and Susan Hume were routed from sleep in their respective rooms by insistent pounding at their doors. "Fire! Fire!" someone screamed.

Bill and Tom are pitchers for the Cincinnati Reds baseball team. Both men and their wives were in town to enjoy a few pleasant days in the city billed as the Entertainment Capital of the World.

"I grabbed my wallet, some pants, and my shoes, "Bill said. "Dona grabbed her purse. I was starting to open the door. We were both being casual. We thought it was probably a small fire."

The Humes knew differently.

"C'mon, you guys!" Susan Hume yelled. "Hurry!"

Both couples rushed for the fire stairs at the end of the twenty-fourth-floor hall and began descending. Bill became separated from the group when he stopped to help a woman try to wake up a friend in a room at the end of the hotel.

After she had gone down several flights of stairs, Dona Bonham realized that she couldn't find her husband. Told by some people that there was no way to get out by going down, she started back upstairs to look for Bill. In the confusion, she lost the Humes.

"I was on the twenty-second floor," she said. "By this time I was light-headed because of the smoke. I thought I was going to be trapped. That's the first time I got scared." A moment later, she found her husband.

"We knelt down, trying to find some air to breathe," Bill said. "We were trying to figure out what to do."

The Bonhams decided to try the roof. After a tense climb they made it. Somewhere in the crowd on the roof were the Humes.

Helicopters were lifting people off, taking women and children first.

"We had to leave the guys," Dona said. "I was really afraid to do that. I knew they would help other people. I told Bill, 'Please don't be the last one down.' "

Several helicopter shuttles later, Bill and Tom were on the ground and reunited with their wives.

Meanwhile, the Krakovers, marooned in their little world, were in their third tense hour of waiting, when a firefighter pounded on their door.

"Get yourselves a wet towel!" he shouted. They did, and Grace Krakover grabbed her purse. Following the firefighter, they went down the firestairs and were outside in fifteen minutes.

The time was 10:15.

Shaken, but physically all right, the couple headed across the street to the Barbary Coast's coffee shop to gather strength.

They were quietly sitting there, sipping on the day's first cup of coffee, when a restaurant employee wordlessly handed Krakover a telephone. It was a reporter from ABC Radio in New York, who had called the coffee shop, tersely asking to speak to a survivor.

"The guy apparently had a scoop because he had the presence of mind to call the hotel across the street," Krakover said. "He interviewed me on the phone, and it played all over the country."

A relative in Miami, a friend in Detroit, and a business partner vacationing in Hawaii all heard Stewart Krakover on the radio that day.

"I had become an instant celebrity," he marveled.

But then, after reviewing their harrowing experience, Krakover was struck by a sudden thought: "Thinking back, we really did something pretty dumb," he said. "Here we were, up there two and a half hours, talking about all kinds of stuff, without ever once figuring out what we were going to do or what we were going to take with us once the firemen got to our room.

"We left a lot of jewelry, some luggage, and my briefcase full of business papers behind. We never thought of taking any of that.

"That part of our brains must have been closed. It was dumb. How dumb could I get? All we ended up taking with us was a bag of dirty laundry!"

Leon Stachowski's former navy service had given him

some experience in handling himself in an emergency. Visiting from Diamond Bar, California, Stachowski and Lori Meisel had lost $250 the night before.

On the twenty-fourth floor, as he and Lori first discovered smoke, he knew they should wet down towels and sheets and line the door. But that helped for only a few minutes; then they were forced out onto their balcony.

"The smoke just rolled out from behind the mirror in the bathroom," said Lori. It was a long wait on the balcony, and Leon considered the possibility of their trying to jump from balcony to balcony to make their way to the ground, but Lori would have nothing to do with it.

Her reluctance probably saved their lives, because few who attempted that kind of thing made it.

With wet towels over their faces, they watched helplessly as a man died on the next balcony. "He didn't have anything over his mouth," Lori said. "His lips turned black, and black blood came from his nose and mouth. He only spoke Spanish, and we couldn't make him understand."

They talked about what they would do if the fire reached them, and decided they'd jump rather than burn to death.

And until they saw the helicopters, they did think they would die. Then, from the casino roof, came a shout from a firefighter that the fire was out and that they should stay calm.

They waited. And they waited.

Finally, after three hours on their balcony, Leon Stachowski and Lori Meisel were conducted down through the fire stairways. As they groped their way down, they saw weak and coughing men and women, some barely able to walk. Those who could walk were carrying those who couldn't. Almost everyone had another person by the hand. Making it to the street with their fellow guests, they hurried to find a telephone and called their families.

"They knew more of the details than we did," Leon said. "They were nearly hysterical. The families of those in the MGM went through a lot that day."

A few hours earlier, Dolores and Roger Mack had begun a four-day celebration of their twenty-fifth anniversary. Now, as smoke poured into their ninth-floor room, they realized with horror that a fire was raging below.

Roger was the manager of an electrical-equipment company in Beaumont, Texas, and a deacon in his church.

Forty-six-year-old Dolores was president of the Ladies' Society of the Saint Pius Church. She had seen death before, close at hand. As a teenager, Dolores had witnessed the devastating explosion of a chemical tanker in the harbor at Texas City, Texas, on April 16, 1947, in which 561 people died. Then, ten years later, she survived Hurricane Audrey, which battered Louisiana and Texas from June 27 to 30, 1957, killing 430.

Now death was near once again. How near, the Macks didn't know, but the screaming sirens and all-out firefighting effort going on below gave them reason enough to fear the worst.

"We're not going to stay here and burn!" Dolores cried to her husband.

Escaping the building seemed the only way to survive. Dolores saw a construction worker's rope ladder suspended near their broken window. She swung herself out on the frail contraption, then lost her grip and fell to the ground, as Roger watched helplessly. Dolores Mack died in her attempt to escape. Her husband made it part way down the precarious rope ladder; then he crashed to the ground. He was alive but suffered two broken legs.

Dr. Charles Smith, a family counselor, and his wife, Ann Smith, were stopping over at the MGM during a cross-country business trip. After spending the day conferring with Jim Reid, known as the Chaplain of the Strip, and several local ministers, they had turned in early. They had a nine-o'clock plane to catch. So at 7:15 Friday morning, Smith closed their suitcases and said he would go down to check out.

"I have two long-distance calls to make," Mrs. Smith said as her husband left. "I'll get them done while you're gone."

She was placing her first call when Smith left the room and headed for the express elevator linking floors seventeen through twenty-six with the casino lobby.

As the high-speed elevator neared the lobby, Smith thought he smelled smoke. When the automatic doors opened on the casino level, he took two steps out into total darkness. He saw nothing—no people, no lights, no furnishings.

He immediately ducked back into the elevator. Someone on the seventeenth floor had apparently pushed the elevator button, and the elevator shot back up. On seventeen, as he emerged, smoke was filling the corridor. Frantic people were dashing toward him, and he warned them not to try to take the elevator down.

He ran into an open guest room and dialed the hotel

operator. One ring. Two rings. Three rings. No answer. He slammed down the phone and ran back into the hall.

Some of the guests raced to the fire exits.

Seeking clear air, he went up the fire stairs, trying each door. None opened until he got to the twenty-first floor. Encountering clouds of black smoke, he started down again.

Far in the distance, he could hear shouts of "Fire! . . . Help us! . . . Get us out of here!"

Sirens were wailing as he reached what he took to be the fifth floor. He kicked out a metal grate and jumped onto the low roof on the south side of the hotel.

Outside, still dozens of feet from the ground, he saw panicky people in the windows above.

People were throwing tables and chairs through the windows. Some tied towels and blankets together and climbed down several floors to the roof he was standing on. With hand signals, he tried to direct the frightened guests to the stairway.

Then he saw the helicopters, and with boyish amazement, he watched them hovering alongside the towering building, their rotor blades beating away the enveloping smoke.

Ann Smith was on the phone to Jackson, Mississippi, talking to Dr. J. Clark Hensley, author of the bestselling how-to book *Coping with Being Single Again*. He was to be the main speaker at a conference she was organizing.

They had spoken for about a minute, when she asked him to hold the line. "I smell smoke," she told Hensley. "I'm going to the window and see where it's coming from."

She pulled open the sliding glass door to the balcony; seeing a little haze in the distance, she thought the smoke must be from a building close by, and she went back to her phone call.

Suddenly there was a haze in the room, and a stronger smell of smoke. She opened the room door and saw a blackening hallway. Running back to the phone, she told Hensley, "There's smoke in the building. I'm on the twenty-fifth floor, and my husband's in the elevator!"

She banged down the receiver and flew to the balcony, where desperate cries reached her ears. A scene of horror had instantly materialized as people smashed windows and flung knotted sheets over the side.

"I barely made it out of the room in time. The smoke was so thick I couldn't see. I looked down and saw two firemen screaming up at us, but I couldn't hear what they were saying.

"The firemen started acting out that we should throw wet towels over our heads and go to a fire exit. I hollered at a few people who were near my balcony that we should meet and join hands and go out together, but I couldn't get any cooperation to do anything collectively."

She tried the hallway again, counting the doors as she groped her way. Almost overcome by smoke, and deciding that it was impossible to make it to a fire exit, she returned to her balcony.

The three firemen were standing where Ann had last seen them. "They kept screaming to us to go into the hallway. We kept trying to yell back that we couldn't."

But their plea seemed so urgent. Thinking that the hallway might be her only chance, not knowing the extent of the fire, she again tried to go out but was again forced back by smoke.

Once more on the balcony, she saw two people fall after attempting a descent down knotted sheets. When the helicopters arrived, she saw their first rescues.

"But they were picking up people one at a time and were so far from where I was, I knew it would be a long time before they got to me."

Physically drained, thinking only of getting to a lower floor, the woman crawled over the railing and suspended herself from the balcony.

"I wanted to make it to the balcony below me, but I wasn't tall enough for my feet to touch the bottom railing. So I decided to hang by one arm and tried swinging out."

Clutching the balcony rail tight with one hand, 250 feet from the ground, she swung like a pendulum out and back until she thought the momentum was enough to drop her safely on the balcony below.

Landing with a crash on the twenty-fourth floor, pain from a broken ankle welling through her body, she rolled out of the way of falling glass. There was nowhere else to go, nothing more she could do.

8 . . .

Emergency Mobilization

When Dr. John Batdorf completed his work at the scene of the fire and returned to Southern Nevada Memorial Hospital, he found things well under control.

In the admitting area, charge nurse April Row was acting as the hospital's traffic cop. "You there. You there. You over there."

That was to have been April's day off. Her husband, a Nevada Test Site employee, had called her just before 8:00 A.M. and told her that she had better get dressed.

Checking with the hospital, April had learned that Civil Defense was declaring the fire a category three. That's serious but not serious enough for the hospital to call in off-duty workers. April decided to go in anyway.

By the time she arrived at the hospital, Civil Defense had upgraded the disaster to a category four. April started calling hospital personnel in.

"A lot of people didn't have to be called," she said. "They heard about the fire and just came in."

The same thing happened at Desert Springs Hospital, where a number of University of Nevada, Las Vegas student nurses had been planning to hold a graduation party. When they heard about the fire, they made their way to the hospital, donned lab coats, and helped in the triage area.

Southern Nevada Memorial was scheduled for its disaster drill that week. "We had a pretty good drill," April said. "It was orderly. There were no people in the halls. The only people who waited were the ones with dog bites." (Metro police dogs, helping with crowd control, had turned on several people who disregarded the cordon.)

On a normal day, seventy-one-year-old volunteer Mickey Gervin was crocheting or making stuffed animals at her post, the information desk near Southern Nevada Memorial's front door. Mickey had been a laundry-room employee at the hospital until she reached retirement age.

There was to be no sewing for Mickey that Friday. When the call came in about the fire, Mickey's first thought was to provide blankets. It was cold that morning, and she knew that many people would leave the smoke-choked hotel in their flimsy nightclothes.

She went down to the laundry room and brought up several baskets of blankets.

"The people who came in were sooty, like they'd been working in a mine," Mickey said. "And they were so quiet. I guess they were in shock."

It was not to be a good day for the elderly or infirm. However, George Kreker, who suffers from emphysema and who was soon to retire from his job as a state official in Illinois, was optimistic.

George had been married to his wife, Sara, for only five days. The Krekers had originally been assigned to a room on the deadly twenty-second floor, but because of a problem with the air conditioning in that room, they had been moved to the seventeenth floor.

The honeymooning couple had originally planned to check out on Thursday but had decided to extend their visit by an extra day. The night before, they had won a hundred dollars at keno.

George reassured his wife. "Don't worry. There's nothing to get worried about. This is a big hotel, with all the fire-protection devices."

He tried lining the door with wet towels, but when smoke still poured into the room, he smashed out a window. All around them, in the hall and in adjoining rooms, people were screaming and yelling. Then George saw the choppers coming in.

At this point, his apprehension increased. He was worried—far more worried than he would admit to Sara.

To the Krekers, and to thousands of other guests awakened by smoke, screams, and terror that day, their luxurious room, with its elegant appointments, in that great, rock-solid giant of a hotel, had turned into what might become a death cell. The reality only minutes earlier had been thick carpeting, a splendid swath of draperies framing gleaming windows, a fine bed with smooth, buttercup-yellow sheets.

Now there were sirens wailing in the street below, choking smoke forcing its way through every crack, and helicopters fluttering madly past the window. This was the new reality, and it was bizarre. If this could be happening at a place like the MGM Grand, nothing made any sense at all.

Two hours after they had heard the first commotion in the hall, deliverance came for the Krekers. At 9:30, a small team of firefighters kicked down the door.

"We've come to take you out," said one. "Be calm— don't panic. You're going to be all right."

Wet towels over their heads, the Krekers were led to a fire exit. Then, holding on to each other, they felt their way down the stairs. His heart was beating irregularly, and gasping for breath, George had to sit down and rest every few flights.

When they reached the sixth floor, he fell into the arms of paramedics, who picked him up and carried him down and out of the building to an ambulance. Only then did he realize just how bad it was.

For two hours, on her preempted twenty-fourth-floor balcony, Ann Smith lay huddled close to the wall. An hour before, she had crawled into the room's bathroom and soaked a towel. The water had gushed out of the faucet boiling hot, but she painfully made her way back to the balcony, cooled the towel, and now was trying to breathe through it, helpless amid the sights and sounds of rescue operations going on around her.

Suddenly there was a shout nearby. A firefighter was calling to her from a balcony three rooms away.

"Are you hurt?" he yelled.

"Yes!" she called back.

"What room are you in?" he shouted. She managed to tell him, and in seconds a pair of firefighters had kicked down the door and were on the balcony.

Using the broken door as a stretcher, they carried her up to the roof, where she was placed in a helicopter and ferried down to an ambulance.

Rumors were the order of the day. With telephone lines jammed with calls, news people were having a difficult time separating fact from fiction.

Everyone in the city had the urge to try to do something to help, and with thousands of organized volunteers meeting the emergency in a hundred different ways, there was a surge by the rest of the public to meet possibly unmet needs.

Thus, when one radio station incorrectly reported that the governor had called on the National Guard to fly in plasma and that there was a desperate need for blood, lines formed minutes later outside the Blood Services Center.

"For us, it was a systems failure," said Ken Reed, executive director of Blood Services. "Of course, we could use the blood, but it really had no direct effect on the fire victims. There was little need for blood in that situation. And blood takes four hours to process. We have to have it on hand before it's needed."

With blood donors lined all the way down the block, Reed finally called radio stations and asked them to tell people that the center had all it could handle.

A benefit was that Blood Services technicians drew 389 units that day, before hundreds of other willing donors went home. The suddenly overflowing blood inventory was shared with other community blood banks around the country.

Every telephone in Las Vegas went crazy that day, particularly the phones in newspaper offices and at radio and television stations.

Information—any information—was at a premium as calls came from all over the world.

At the *Sun,* assistant to the publisher Ruthe Deskin instructed classified advertising personnel to help field the hundreds of phone calls pouring in, to relieve the busy newsroom staff.

By midmorning, the confirmed death toll had reached twenty-six, and the numbers were still mounting.

Still in his van, blocked by emergency vehicles, 007 was broadcasting live to the San Francisco Bay area. As he reported developments, other reporters borrowed his car phone.

At KORK radio, Andrea Boggs, who would broadcast for nine straight hours, was infuriated with some calls from the national press.

"They asked me if bets were being placed on the number of bodies they'd find. I didn't lose my temper. I'm very proud that I didn't lose my temper."

Her student broadcasters toiled by her side, and she thought, *The things I've been teaching my classes, I'm applying. You do have to think fast and do seven things at once.*

When the death toll rose higher, she fought back tears. There was simply no time to cry.

*   *   *

Watching the helicopters from the low roof he had made his way to, Dr. Charles Smith was thinking, *They're flying so close to the building. The rotor blades will whip into the windows.* But they didn't.

Now he had to figure out how to get off the roof, and after a few more deep breaths of outside air, he plunged back into the hotel staircase. The door down to the next floor had been locked when he tried it before. Now it was open, and he scrambled down the darkened stairwell and pushed open a door at the bottom. He was out.

Men in yellow jackets were running past. A tangled skein of swollen fire hoses covered the pavement. There was the steady hum of voices all around, blending in with the rumble of generators and helicopter engines.

He heard someone say they were taking many of the guests to the swimming-pool area, and he headed that way. Ann wasn't there. Then he ran to the area where the helicopters were landing, but he was held back. With the whirling chopper blades splitting the air, bystanders had to be kept at a distance.

Plunging into the milling crowd, looking into every blackened face, Smith could not find his wife. Then he went to where the schoolbuses were lining up and watched them take on evacuees togged out in odd bits of clothing, or wrapped in sheets and blankets. Drained, suddenly indecisive, he stood for minutes just staring at the departing buses.

Going to the Convention Center seemed to be the sensible thing to do. Ann might be there already, or someone might have information about her. As the next bus prepared to leave, he sprang aboard.

The Las Vegas Convention Center, with 785,000 square feet of floor space, is the largest one-level convention facility in the world. It is big enough for a convention or two, a county fair, a rodeo, a carnival, and university graduation exercises to go on simultaneously, without using half the building's capacity.

On Friday, November 21, the enormous hall, for the first time in its two-decade history, was being used for an alternate purpose—it was the center of disaster operations.

By eight o'clock that morning, the Red Cross was ready. The six workers first on the scene had cots set up, and oxygen and other equipment were on hand.

After alerting other Red Cross workers, Betty Bres had raced to the facility. When the first schoolbus filled with evacuated MGM Grand guests arrived on the scene, nurses and doctors were reporting for duty.

Their patients were the walking wounded. Hundreds of cases of cuts and smoke inhalation were treated on the spot; the more serious cases were sped to hospitals.

Language interpreters were desperately needed, and a call went to radio stations to broadcast an appeal for them.

In the disaster plan, the role assigned to the Red Cross is communication between victims and families. In the chaos of getting people out of the hotel, couples were separated and families and friends broken up.

The first concern of the confused and frightened people pouring into the rescue center was to locate relatives. As Central Telephone-Nevada (Centel) crews installed banks of special telephones, and volunteers arrived to answer inquiries, the resources of the Comdex Exposition were also brought into play.

Computers from the giant exhibit in a nearby hall were moved into the rescue area, and computer operators among the convention delegates and exhibitors began to compile endless lists. As people arrived and registered with the Red Cross, their names, room numbers, and home cities went into the computer.

"We encouraged people to stay put," Betty said, "and let us look for their friends or family members. We'd get a description and go looking. I'd walk up to somebody and say, 'Are you Señor Rodríguez?' And when he grinned from ear to ear, I knew I'd found him, and I'd take him back to his wife."

For the families of the injured, it was not always so easy. Charles Smith arrived at the Convention Center at eleven o'clock and registered with the Red Cross. They had no record of Ann. He waited ninety minutes, then checked the first computer lists just coming out. Her name wasn't on them. Then a volunteer started calling hospitals for him. When she reached Sunrise Hospital, the answer was, "Yes, we have an Ann Smith here. She'll be all right."

Ann was less than a mile away. The Smiths were reunited at 1:30 in the afternoon, six hours after Charles stepped out of that elevator into a well of darkness.

Over and over that day, the special needs of the emergency were met. It was a strange and fortunate twist of fate that the helicopters of Operation Red Flag were nearby, that the computer convention was in progress, and that more than a thousand doctors were among convention groups at the MGM.

"Whatever we needed just seemed to appear like magic,"

said volunteer Bonnie Gragson. A licensed practical nurse, wife of former longtime Las Vegas mayor Oran Gragson, she had worked all morning giving oxygen and helping with survivor registration.

"During that time I didn't even look up. When I did, I couldn't believe it. There were tables of clothes. The hotels had sent food. Blankets came."

Working inside the Convention Center, not stopping for a minute, neither Mrs. Gragson nor others there understood to what extent the community was mobilizing.

The clothing Salvation Army Major Rex McCulley had hastened to get was in orderly piles on long tables, neatly organized by sex and size. Men, women, and children clutched blankets around themselves while they waited their turn at the table. Prosperous hotel guests, accustomed to only the best, were grateful for someone's castoff shirt and jeans.

The grim day had its funny side to Dr. Abe Sosman, a physician from Milwaukee.

"People told us not to come to Las Vegas because we'd lose our shirts here," he joked. And pointing to the pajamas he had been wearing when he fled the hotel, he added, "I sure lost mine!"

A Laredo, Texas, man stopped the busy Major McCulley to say, "I'm a Rotarian, and every Christmas season I'm out there ringing the bell for the Salvation Army. For the first time, I understand why I'm doing it. You can bet I'll be out there again this Christmas."

It was a polyglot gathering of United Nations proportions; people from Mexico, Spain, France, England, Germany, Argentina, Switzerland, and Japan boiled through the hall. As interpreters arrived, they set up information posts with large overhead signs. Other interpreters, designated by badges, circulated through the crowds, while loudspeakers blared announcements in several languages.

University of Nevada, Las Vegas professor Claude Rand came to help French-speaking people and also found himself broadcasting for an hour in French over Radio Luxembourg.

Among those aiding Spanish-speaking fire victims was Nestor Miller, who had been at his desk in a Las Vegas travel agency when the fire news began to spread. He went to the Convention Center and helped interpret for doctors; then he went from person to person, offering what assistance he could.

A Mexican woman clutched at him, sobbing that she couldn't find her daughter and son-in-law. "How many are dead?" she cried as Miller tried to comfort her.

He said, "It's not very many, and they're all Americans."

That wasn't the truth. About thirty were known dead by that point, and many of them were Mexicans, but he thought it was better right then not to tell her that.

The fate of the 375 Mexican guests known to be at the MGM Grand prompted hundreds of phone calls from Mexico. Four of them came from Mexico's president, José López Portillo, who was seeking information on the whereabouts of many friends. The confusion was compounded by language problems and because many names were incompletely listed. Mexican names include both the mother's and father's surnames, and Red Cross workers and other Convention Center registrars did not always understand this.

The inquiries from the presidential palace in Mexico City yielded some information, but none about the president's closest friends involved in the disaster, Fernando Lobo Morales and Susanna de Lobo, a couple from Monterrey. As the hours wore on, their fate was still undetermined.

Nothing in the Convention Center could dispel the gloom of the day. The number of dead escalated with each news report, and rumors circulated around the hall that the toll might go over one hundred.

Survivors streaming in had horrifying stories to relate. Major McCulley talked to a man who had been on the ground floor of the hotel. When the casino exploded in flames, he was blown right out onto the street. Miraculously, he wasn't badly hurt.

One Salvation Army worker was a former mortician. After several hours assisting the living, he went to the coroner's office to help identify the dead.

A buzz of recognition ran through the crowd in the hall as Evangelist Billy Graham appeared. A young attorney he stopped to talk with told him that his three-month old baby was missing. A fireman had taken the child from the building along with other children, and they had not yet located the youngster.

Graham prayed with many still looking for their families, and when 007 appeared at his side, the religious leader said over the radio: "This is the greatest tragedy Las Vegas has seen in its history, and it's unbelievable. . . . I've met some people here from Ireland, from England, from Australia. I met a lawyer from Florida. A medic took him away. He'd lost his wife. He saw her lying on the twelfth floor and he hasn't seen her since. . . . What I've tried to do is have a little prayer with some of them and be sort of a chaplain to them."

The army of Convention Center workers was augmented by a second army fanning out through Las Vegas streets to provide transportation of people and relief supplies. The Jaycees turned up with vans to shuttle victims from the MGM Grand to the disaster center and from there to hotels and motels. One Jaycee, who had no suitable vehicle, rented a van for $150. Sam Boyd of Sam's Town Hotel and Casino offered to pay for gas, for the Jaycees and other private-car owners serving as drivers.

While they labored, Convention Center workers were appalled by things they kept hearing. Bonnie Gragson wiped away tears as a volunteer related what her son, a Metro policeman, had seen that morning. "He told her he'd pulled plenty of people out of wrecks, but he'd never seen anything like this. He had gone into a stairwell at the hotel. People were sitting there on the floor with their eyes wide open. He went over to help them up, and when he touched them, they toppled over. They were all dead."

9 . . .

Last Rites

Twenty-year-old Luke Carr, a lifeguard at the hotel, was getting the pool ready for the day's swimmers and sun-bathers when the first people spilled out of the smoke-choked resort.

Making use of the wheelchair kept in the pool house, and grabbing an oxygen kit, Carr moved quickly to treat victims of smoke inhalation. Assisted by uninjured guests and hotel employees, he started helping people cut by flying glass.

The pool area soon became a staging ground for ambulance crews picking up hospital-bound fire victims. The schoolbuses pulled up nearby.

Carr worked with ambulance attendants and paramedics to lay the injured on flat lounge chairs or on the grass until they could be given emergency first aid or rushed to the hospital.

He offered his help to a fire captain, who needed some-

116

one familiar with the hotel's layout so firefighters could carry out rescue efforts.

His first assignment was to go up the stairway to the seventeenth floor to search each room for survivors and bodies. Carr then led the firemen to the rear service elevators.

With a crowbar, they forced open the elevator doors. It was young Carr's first glimpse of death.

Two people were dead in the first elevator the firefighters opened. Two more were found in the second elevator, and three others had succumbed in the third.

Carr was sick.

"We need more men on the roof," squawked the captain's two-way radio. Carr was lifted there by helicopter.

When the chopper touched down atop the hotel, a firefighter ran toward him, calling, "We need you to carry 1911s."

"1911s?" Carr asked. "What's that?"

"Dead bodies," was the reply.

Until that day, the closest Carr had ever come to seeing death was giving a drowned girl mouth-to-mouth resuscitation until paramedics arrived to revive her.

Now the appalling sight of the dead and dying would be forever etched in his memory. He broke out in a cold sweat and felt his legs giving out under him.

"Luke, you're too young to let this get to you," the captain said. "It's a job. It's got to be done, and you just keep doing it."

Carr didn't feel any better. His insides were turning at the sight of the bodies in the hallway.

"Just put it out of your mind," the captain told him. "It has to be done. We need to keep going."

"How do you put it out of your mind?" Carr asked. "I *can't* put it out of my mind."

"You will," the captain replied.

Also called into action that morning was Las Vegas acrobat Frank D'Agostino.

As the "black and bad" smoke spilled over into hallways and guestrooms, D'Agostino scrambled to bring people to safety.

He made seven or eight trips—he lost track of how many—to the deadly twenty-third floor. Then he was overcome by smoke.

D'Agostino would wake up later that day in Desert Springs Hospital.

If the rooftop scene was Vietnam, the medical command post at the rear of the hotel near the helicopter landing area resembled Korea—a scene from *M*A*S*H*. As the choppers landed, medics and volunteers rushed under their whirling blades to assist people off, then helped them to impromptu pallets made of sheets and towels spread out on the parking lot.

Helping calm fire victims, assisting them as they came off ladders and out of helicopters, were two seventeen-year old seniors from Clark High School, who had been in the school cafeteria at 7:30 that morning when they heard about the castastrophe at the MGM Grand.

Best friends, co-captains of the school drill team, GiNié

Wilson and Laura Fielden had impulsively driven to the hotel.

"People will be frightened and hurt, and we might be able to help them," they said to each other. So the A students played hookey for the first time in their lives. They arrived at the south side of the hotel shortly before 8:00 A.M.

They stared with horror at the burning building and the terrified people waving from windows; then they were quickly absorbed into the rescue team when a firefighter ran over and thrust a load of bedsheets and blankets at them.

"Here, start spreading these on the ground," he commanded, then hurried away.

As half-clad people stumbled into their arms, the teenagers, and soon other passersby, sped them to waiting medics and EMTs.

Many of the guests escaping from the south side of the hotel were Mexican, and the girls' high school Spanish made communication possible.

"You're going to be all right," they would say in Spanish. The fleeing guests were crying and screaming. And they were cold. GiNié and Laura wrapped blankets around shivering bodies and hugged them with their strong young arms. One blanketed man had fled the hotel in the chill morning air wearing only cowboy boots and undershorts. A little boy sobbed as if his heart would break, and husbands and wives were calling for each other. A woman they tended had blood pouring out of her leg, but she kept saying, "Never mind me—where's my husband?"

They conducted people off ladders, then made dozens of runs to help people out of helicopters and into the hands of paramedics.

The thousands of fleeing guests were given oxygen and a quick checkover. As medical examinations were made, those not needing treatment were conducted onto waiting schoolbuses, to be taken to the Convention Center. The injured, loaded into ambulances, were sped to hospitals.

Filthy with soot, impervious to fatigue, the teenagers ran back and forth for almost four hours. In the course of the morning, they watched three people die. Then, tears streaming down their smoke-blackened faces, they reached out for the people still pouring out of the hotel.

By 9:30, the holocaust in the casino was under control. Some small fires still burned, so after a five-minute break, Captain Rex Smith went back into action.

"The draft from the door kept the smoke back about fifteen feet from the door, and we were able to go in without our air packs at first," Smith reported. "Later we went back outside, grabbed an air pack, and took a team of firefighters inside to smother the last flames."

The firefighters, whose yellow jackets and pants were caked with ash, walked along the corridor hugging the north-east side of the casino. It was like hiking through a marsh.

The devastated casino, Smith said, "give me kind of an unrealistic impression. We'd been in the MGM several times on inspection, and now it seemed so different.

"Melted pools of plastic were everywhere. It all made a

picture of a large and devastating kind of fire you hope you never have to be in.

"When we reached the elevators, we saw two bodies sprawled on the floor. They had apparently taken several steps into the casino before they realized they were walking into an inferno. We saw there was nothing we could do for them. They were badly burned and fairly unrecognizable."

To *Sun* police reporter Gary Gerard, the scene inside the casino was like pictures of bombed-out buildings in World War II.

In his words, "A city fire captain advised [the press] to wait a little while before going in because there was still the possibility of cyanide gas in the casino. On my first trip into the area, I walked into the lobby about as far as the stairs leading to the casino pit. The place was demolished. On my right, where the keno lounge was, the ceiling had collapsed. Beams of light filtered through the steamy haze. The chairs in the keno area were burned, but I saw no people there. In the main casino pit, the floor was awash with several feet of water, and water was coming down from the ceiling, as firefighters continued to pour thousands more gallons of the stuff into the casino to make sure every last glowing ember was extinguished. It was like walking through a rainstorm."

Gerard retreated from the scene to borrow some boots, then returned to the casino with a Clark County firefighter.

This time he got as far as the registration desk on the northwest end of the lobby. From that vantage point, he saw what appeared to be the forms of two, perhaps three, people, slumped over their chairs in the small piano bar that straddles the pit. "I said 'forms,' because they were just that. They seemed so unreal." There were moments in there when he thought he was just looking at a violently contorted fragment of something that had burned in the fire and saw in its shape the illusion of a body. But these were not illusions.

In the distance, silhouetted against the eerie glow of flashlights shining through the cloud of steam, he saw—or thought he saw—the figures of two people standing near slot machines. He wasn't close enough to make out facial features; he wasn't close enough to touch them to see if they were real. *Were* they real? the awful truth was—yes.

It was cold and damp in there. The stench was sickening. He went outside and, looking up, saw an air-force Jolly Green Giant helicopter hanging motionless above the showroom wing of the building. Flames were still flickering in the pavement near the main entrance. The front canopy had been destroyed, and there was fear it might collapse. Security guards warned, "Get away from there."

At Holy Family Church, a few miles east of the Strip, the Reverend Glenn Smith, a visiting priest from Saints Peter and Paul Catholic Church in Cary, Illinois, heard about the fire as he was saying mass. As soon as services were over, he drove directly to the hotel, pointing at his collar to get through the police cordon. Someone had to minister to the dead and dying.

As the priest told the *Sun*'s Elliot Krane: "I went to the back of the hotel, where the helicopter-landing strip had been set up, and decided to try the staircase. The fire department had smashed the locks on the doors to each floor, but it was pitch dark in the stairwell. Firemen directed me to some people who had died on the fifth floor, and then we went up together to the fourteenth floor, where there were more bodies. I assumed that the people had died of heart attacks, but there were so many dead that I realized it was something else. By the time we reached the nineteenth floor, we found that most of the people were either dead or dying. With the windows smashed in the room, there was a cross-draft, which helped us to see down the halls."

As the priest plodded through the hotel, he could hear the announcements from helicopter bullhorns urging people to stay in their rooms. In his rounds, the Reverend Smith was finding only dead bodies:

"They seemed to have accepted their fate," he said. "One woman wrote HELP in foot-high letters in lipstick on the window before she died. One husband had wrapped his wife in a wet bedsheet and wet towels and then slumped in a chair by the window. When I got to their room, they were both dead."

On the twenty-first floor, he found the body of room-service waiter John Ashton, his breakfast tray nearby. He saw a husband and wife hugging each other—in their final embrace. Another couple had died while in prayer, and one man was fatally overcome with a Gideon Bible in his hand.

In one room, on the dresser, the priest saw a picture of a couple—a picture apparently taken the night before, as they were celebrating an anniversary or birthday in the showroom. Now the man and woman in the photograph were dead at the foot of the bed.

Smith was joined by another priest, the Reverend John McShane, and together they anointed the dead and dying until warned by firefighters that it was dangerous for them to remain in the stricken hotel.

Impersonally, death visited the old, the young, and the middle-aged that day. Those unlucky enough to be on the ground floor when the fire swept through were among the first victims.

Genell McDowell, from Memphis, Gustave Guidry, from Larose, Louisiana, and slot-machine mechanic Mark Hicks were killed in their tracks near the registration desk.

José Vásquez, twenty-five, credit manager for a Mexico City pharmacy, was slumped to the floor near the middle of the casino.

Twenty-year-old Phyllis Thomas, who had grabbed the cash box and run as she was told to do, stopped to help an elderly hostess who had fallen. Unable to pick the woman up alone, she called to security guards for help.

The older woman was carried out. Phyllis's body was found between the registration desk and the side entrance.

Jac Keller, sixty-one, and his wife, Blanche Keller, sixty-four, from Carmel, Indiana, were overcome in the casino elevator hall. They had apparently decided to spend a little time gambling before their scheduled 10:10 A.M. Friday departure flight.

David Sanders, thirty-nine, who owned an office-supply store in Indianapolis; his wife, Barbara, thirty-seven, a registered nurse; David Asher, Jr., thirty-nine, an assembler for Chrysler; Karen Andrews, thirty-six, of Indianapolis; and John Monaweck, fifty-six, from Little Rock, died as their elevators opened on the casino level.

Two MGM Grand security guards, thirty-seven-year-old Joseph Hudgins and fifty-three-year-old Sherman Pickett, as well as William Gerbosi, twenty-four, from Western Springs, Illinois, and Dr. David Potter, twenty-four, from La Grange, Illinois, were overcome by smoke in the Ritz Room, part of the hotel's convention enclave, east of the lobby.

Elevators, and locations near elevators, were especially deadly. The graceful, plushy appointed elevator lobbies held fifteen smoke victims. Found in the twenty-fifth-floor lobby were Dellum Hanks, forty-eight, of Euclid, Ohio; Barbara Middleton, thirty-nine, of La Mesa, California; and MGM Grand maid Willie Lee Duncan, fifty-five. In the twenty-fourth-floor lobby, where they had fallen while attempting to escape, were Daniel Peha, twenty-three, of Mexico City; Thomas Spagnola, forty, of Des Moines, Iowa, and Diane Pangburn, twenty-three, also of Des Moines; and two visitors from Vail, Colorado, Steven Holschuh, thirty, and Catherine Sanders, twenty-three.

Young Pablo Pedro Sierra García, twenty-two, of Mexico City, was found near the body of his mother, Margarita García de Sierra, forty-three, in the twenty-fourth-floor lobby. His father, Manuel Sierra de la Visitación, also forty-three, fell just inside the door of Room 2403, only a few steps from the elevators. In the twentieth-floor elevator lobby were discovered four bodies, those of Barbara and Allan Soshnik, thirty and thirty-one, a popular young couple from Atlanta, Georgia, and Laura and Victor Castelazo, thirty-three and thirty-five, from Aguascalientes, Mexico.

The deadly smoke, racing up elevator shafts, overcame Elmira and John McQuithy, of Marion, Indiana, who had come to Las Vegas every Thanksgiving for six years. Elmira was seventy; John, sixty-three. They were found in an elevator stopped at the twentieth floor, along with David Blair, Jr., twenty-six, of Columbus, Ohio, and James E. Thebeault, thirty-two, of Mansfield, Ohio.

Stairwells, meant to save lives, were often death traps instead. Eight people succumbed in stairwells: twenty-three-year old Roberta Peterson, of Chicago; Patricia Louise Tunis, sixty, of North Hollywood, California; Dr. Donald Nilssen, fifty-nine, and his wife, Janet, also fifty-nine, of Omaha; hotel maid Elizabeth Barresi, fifty-three; Donald M. Shaffer, forty-four, of Morgantown, West Virginia; and Raphael Iadeluca, forty-five, of Montreal, Canada, who was found on his knees, bending protectively over the body of his wife, thirty-three-year-old Angela Iadeluca.

Often just proximity to the elevator banks was enough to bring swift unconsciousness and death. Lori Ann Nose, nineteen, and Carol Ann Mayer, thirty-six, both of Parma, Ohio, died inside Room 2501. Richard O. Johnson, forty-one, of Bloomington, Minnesota, succumbed inside Room 2302; Edward Herring, forty-six, and his wife, Genese, thirty-five, from Irvine, California, were found in Room 2102; Mary Ann Vassoughi, forty-one, and her husband, Dr. Houshang Vassoughi, forty-four, from Lower Burrell, Pennsylvania, died in Room 2001; and the Littmans, Ellis C., sixty-nine, and Roslyn, sixty-three, of Frontenac, Missouri, perished in Room 1902.

It was an irony that day that the higher the floor and the more convenient a room's location to the elevators, the greater the chance of death.

10 . . .

The Grim Search

By two o'clock Friday afternoon, the news of the MGM Grand blaze had circumnavigated the globe.

Oceans of ink and countless hours of broadcast time had already been expended in describing the events of the day. A swarm of print and broadcast journalists assembled near the Flamingo Road fire-department command post for an impromptu press conference called by Clark County Fire Chief Roy Parrish.

Parrish and Captain Ralph Dinsman, the department's information officer, were among the most sought-after persons at the fire scene. They were the ones with the latest update on the firefighting effort and the number of confirmed dead.

They helped set up interviews with firefighters and arranged for access to the ruined underbelly of the hotel. But to many of the national reporters, Parrish and Dinsman were men who "tried to lead you around the bush" and "avoid the hard-line questions." One newspaper reporter from Los Angeles complained that "regular" firefighters wouldn't talk to him. "You've got to get the chief's okay."

Despite these reservations, the herd of reporters flocked around Parrish as if pulled in by a powerful magnet.

"We received the report on the fire at about 7:15 this morning," Parrish said as reporters, who had by now formed three tightly knit concentric circles around him, poked microphones and camera lenses at him.

"The first units arrived within a minute. The men were responding to a fire in the kitchen area. There had been rumors that the fire began in the basement and at the new construction site. To the best of our knowledge, these reports are false."

Some newsmen had been pushed to the outer edges of the huddle and caught a sentence here and a sentence there.

Las Vegas Sun reporter Gary Gerard, who had muscled his way in and was pushed out several times, strained to hear what was being said.

"The fire appeared to have broken through the kitchen ceiling, rolled along the twelve-foot open space between the casino and the roof area, and crashed through the false ceiling into the casino area," Parrish said. "A wall of fire fell down."

Next, he spoke of the death count. When Gerard made his first phone call to the *Sun* that morning, there had been two confirmed dead.

At 8:20, there were three. Shortly after 9:00, Dinsman told reporters that twelve had died, most of them in the casino. An hour later, the tally marks on county Deputy

A frightened survivor gazes up at the burning hotel.

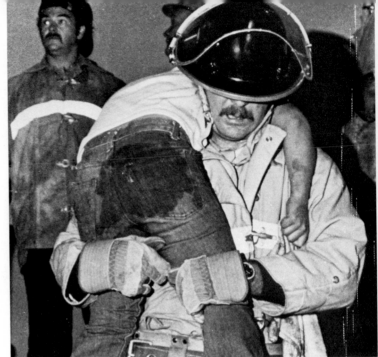

An unconscious little boy is carried out by a fireman. The boy later recovered, uninjured. (UPI photo)

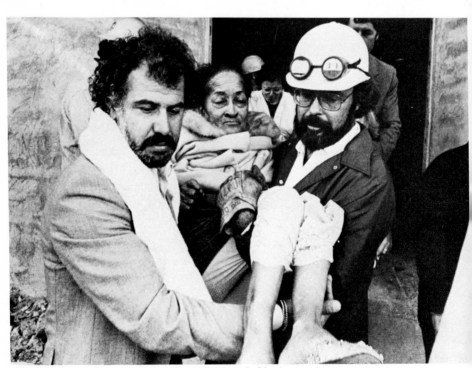

An elderly survivor is carried from the hotel. She is wearing a coat over her nightwear. (UPI photo)

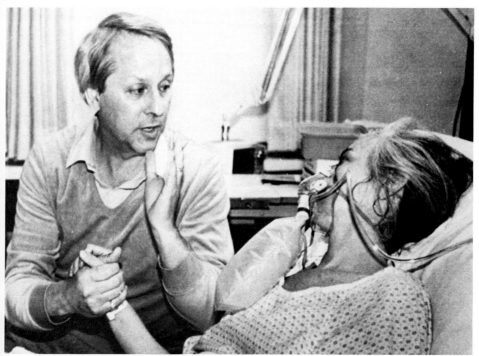

Dr. Charles Smith of Nashville, Tennessee, visits with his wife, Ann, in her bed at Sunrise Hospital in Las Vegas, where she is being treated for smoke inhalation and a broken leg. (UPI photo)

section of the below-casino-level grand arcade. ...te the shattered glass panels advertising the ...GM Grand Theatre and the Jai-Alai fronton.

Burned-out and gutted slot machines in the casino.

Ashes and rubble under the front entrance canopy. The gutted automobile is in the background, where firemen continue to hose down the smoldering ruins.

The Salvation Army, the Red Cross and local residents supplied clothing, shoes and food for the fire victims. Besides being a medical base, the Convention Center was used for distributing clothing and feeding the hotel guests. (UPI photo)

Aftermath press conference: Mike Patterson (left), head of CCFD investigation division and CCFD Chief Roy Parrish.

Nevada Governor Robert List arrives to survey the damage. Bruce Spaulding, Clark County manager, is at his right in firecoat.

Reporters converge at the Benninger Benedict-Rothkopf press conference

The top officials of the MGM Grand Hotel at the weekend press conference. From left: Al Benedict, chairman of the board of the hotel; Fred Benninger, chairman of the board of the corporation; and Bernard Rothkopf, president of the hotel.

nside the casino. Note the missing
eiling panels, the stripped chandeliers
nd the burned-out slots.

The discolored statue of Neptune continues
to stand guard over the front entrance lagoon
under the burned-out canopy.

Main entrance after the fire. Note the hoses snaking their way into the casino.

Kirk Kerkorian in happier days.

The new casino, 150 yards long, which reopened the hotel on July 30, 1981.

Fire Chief John Pappageorge's yellow legal pad had added up to thirty-seven.

"We have seventy-five confirmed dead," Parrish was now saying. "Ten of them were found in the casino area."

Then, a barrage of questions:

Q: Why were there no alarms sounded?

A: The hotel has a manual alarm system, apparently hooked up in the basement. We haven't checked yet to see why it didn't go off.

Q: Did the hotel have smoke alarms?

A: There are none in the structure.

Q: How many were injured?

A: There are at least three hundred, so far. Most of them were taken to Desert Springs, Southern Nevada Memorial, and Sunrise hospitals. Many were taken by bus to the Convention Center, where we will be getting their names.

Q: What was the extent of the damage?

A: The fire was held in check in the shopping arcade when the sprinkler system went off. We've got mostly water damage there. The showrooms and some of the restaurants, except for the Deli, were spared.

Q: How many people were in the building at the time?

A: We figure there were about forty-five hundred to five thousand people in the highrise portion of the hotel. With the employees and the people in the casino added on, I'd say about eight thousand all together.

Q: Is there any reason to think the fire was suspicious?

A: I think we have ruled out arson.

*   *   *

More questions, concerning the building and fire codes in Clark County, and the outdoor press conference was over.

Reporters dashed for telephones. Every television network newscast would lead with the MGM Grand fire story that night. Editors in New York scanned film for TV special reports. And in newspaper offices around the world, the story was being readied for Page 1.

Later, Ralph Dinsman explained that the fire had come very close to entering the high-rise. That would have meant a death toll in the thousands. If firefighters had not succeeded in controlling the heat buildup in the elevator shafts, there would have been another uncontrollable wall of fire. Dinsman said that a flashover in the elevator shaft could have happened at "any time."

It may never be known how close it came to having an inconceivable loss of life at the MGM Grand. Some firemen have confided, however, that this was just seconds away.

By 2:30 Friday afternoon, people who had volunteered to search for bodies were forming teams on Flamingo Road.

"They issued us full fire-department turnout gear, and we went in," said John Jones. He had put 117 miles on his ambulance shuttling the injured and sick to nearby Desert Springs and Sunrise hospitals since 8:30 that morning.

John was part of the second rescue team to enter the resort. He was accompanied by a fellow SNVFR volunteer, Eric Hutt, several local fire-fighters, a man from the Clark County coroner's office, an American Ambulance driver, and an employee of the Nevada Medi-Car, a nonemergency transport service.

The search team entered the casino through the main entrance doors.

As he made his way through the dark and dripping building, where death stalked every corner, John thought, *This is not real.* It was almost impossible to link the blackened wasteland at his feet with reality.

A stench filled the air. Wisps of steam rose from an ocean of water that flooded the casino floor. Gnarled ceiling supports and huge fiberglass ducts sagged from the ceiling. A waterfall was rushing down.

In the distance, John could see the yellowish-white flicker of flashlights piercing the haze. It was cold. The devastation was complete.

As Jones marched along, he saw forms of charred bodies. One was sitting at a 21 table. Another was at a bar. Some were standing in front of slot machines, which now resembled mechanical tombstones.

To deal with so much death at one time was an enormous emotional drain.

"But then you say to yourself: 'Now comes the hard work. You've got to do the job you came to do.' And you try to block it out," John said later.

He remembered carrying six bodies, as rigid as department-store mannequins, to the coroner's van. Before the coroner's people left the front of the building, they had thirteen bodies.

John's search team was then sent to check the nineteenth through twenty-sixth floors for more bodies.

Toward the end of the third hour, he was becoming

weary. He wasn't physically sick, but his legs were hurting, and his whole body was in pain.

I can't go up there again, he told himself. *I can't pick up one more body.*

Renée Jones, who had been assigned to operate the first-aid station set up in the southeast parking lot, was wondering where her husband was. He had kissed her goodbye at eight o'clock and sped to the MGM Grand in his station wagon.

She was now part of the body detail, along with Susan McManus, the wife of SNVFR president Shaun McManus. They were waiting for the dead being lifted off the roof by air-force and civilian helicopters.

"Sue had never really been confronted with death," Mrs. Jones recalled. "When the choppers brought down the first group of six bodies—who were carried off by stretcher and placed on the pavement next to the body bags—she burst into tears.

"But she regained her composure and worked like a trooper the rest of the day."

Many of the dead, Mrs. Jones said, "looked like you could give them a little oxygen and they'd be all right."

One man was brought out of the hotel with his reading glasses on his forehead, holding a book in his hands with a finger saving his place. Renée Jones presumed that he had been overcome by carbon monoxide, which he couldn't see or smell.

She, too, reached her breaking point.

"Crazy things go through your mind," she told a reporter. "There was this air-force man who was wearing a head set and a silver fire suit near where I was working.

"I just wanted to know when it's going to stop. The bodies kept coming and coming, and I wanted him to tell me when they're going to stop.

"He wouldn't answer me. I felt the panic rising inside me. Then he grabbed me with his big bear hug, and all I saw was silver. I took a deep breath and calmed down. There was something comforting about being smothered by that silver fire suit."

She went back to work.

"I've grieved for every person I ever lost. But most of the people I treated before were sick, and you try to comfort them. But the people they were bringing down off that roof—I couldn't accept that. They've got things to do. They've got things waiting for them out in the world. And it was all taken away. All those beautiful people with so much to live for.

"Some of the firemen were weeping as they came out of that building. But as soon as they put a new air pack on, they went back inside again. I never saw a fireman cry before."

Hours after the tragedy, the fate of many missing people was still undetermined. Dr. Stuart Berger, on duty at Desert Springs throughout the day, knew that his visiting mother and sister had been on the fifteenth floor of the MGM Grand. Five hours after the fire began, he was told that they were all right.

Robert Bobbett, R.N., was on duty in the Desert Springs emergency room that morning and had ample opportunity to talk with many of the people who fled the scene of the MGM Grand fire.

Most of their comments were laudatory. A handful were not.

"Some patients complained of the lack of organization at the fire scene," Bobbett said. "They said many of the victims panicked. Some could not believe the way people were behaving out there. 'They were acting like animals,' one woman told me."

Elsewhere in the hospital, a little girl was looking for her mother.

She sat quietly in the hospital lobby until one of the nurses found time to lead her through. "She was very stoic. She didn't shed a tear," a nurse said. "Then, when she saw that her mother was alive, she broke into tears."

Patients with windows facing the Las Vegas Strip cherished those rare moments when the scream of ambulance sirens subsided.

Maybe that's the last one, they hoped.

A few minutes of precious silence. Then another siren.

Please, God—no more.

At Desert Springs Hospital, day nursing supervisor Virginia Empringham had a call from a frantic husband asking if Desert Springs had any Jane Does on its patient roster.

The man said he had been told that his wife was dead. But he didn't believe it. He told Empringham, "I've called all around. You must have a Jane Doe there."

Empringham replied, "I am sorry. All our patients have been identified."

In the aftermath of the fire and their escapes from the hotel, people had just one thought: to go home. Often without wallets, tickets, or luggage, many scantily dressed, they flocked to McCarran International Airport.

At the airport, as MGM Grand guests poured in, passenger agents swiftly reissued tickets to speed guests on their way.

Delta Airlines salesman Mike Calmy said, "We routed them to other lines—anything, just to help them get out of here. They'd had enough trouble."

Waiting for Delta's Flight 912, scheduled to depart for Dallas–Fort Worth at 1:55 P.M., about twenty disheveled people compared notes on their experiences.

One couple were honeymooners. He wore slacks and a corduroy jacket but no shirt. Both were shod in green hospital slippers.

"I lost my wedding ring," the bride said. "But we're lucky—we're safe."

Another woman, still blackened from the smoke, worried about her hands.

"I've washed them over and over, but I can't get them clean."

On the two-hour flight, some somberly stared into space, and others chattered nervously.

A trio of young girls changed into bright LAS VEGAS T-shirts. One said, "I can't believe the things I've seen today. We never heard a warning. No one let us know there was a fire."

When the plane landed at Dallas–Fort Worth airport, television newsmen were ready with cameras, as fire evacuees rushed into the arms of family and friends.

*　　*　　*

On the gutted ground floor of the hotel, Nevada Governor Robert List, wearing a hard hat and a firefighter's jacket, plodded stonily through the ruins. He was accompanied by Al Benedict and Fred Benninger, the two highest hotel officials, who had sped back from their fishing trip when informed of the disaster.

Emerging from the casino, Governor List spoke to a battery of cameras and reporters. "This is a haunting shell of a firestorm," he said. "It's black, ugly, and sickening." Benninger and Benedict, the governor said, "were stunned, just stunned. In fact, flabbergasted."

List announced the formation of a blue-ribbon panel to "examine our ordinances and codes in Nevada and compare them to national standards.

"I want to satisfy myself that we are doing everything we can to minimize the chances of a recurrence of the MGM tragedy."

He said he would assemble the best experts he could find, both within and outside the state.

The saddened governor pledged help in expediting unemployment-insurance payments for the forty-five hundred MGM Grand employees thrown out of work by the disaster.

List had first visited the Convention Center and a hospital. He spoke pensively of talking with some of the victims.

While mourning the loss of life, he compared the toll with the number of potential fatalities, since there had been eight thousand people in the hotel when the fire broke out. That fatalities were not greater, he said, was "a testimonial to our fire department and the hotel staff."

Still numbed by the devastation inside, List said that it

had hit him on Thursday night that there were thousands of people enjoying themselves amid the color and music of the MGM Grand. The next morning, "There was no color or sound . . . just dripping water."

Southern Nevada Memorial Hospital, in accordance with disaster procedures, discharged patients who were not in serious condition, in order to free beds for the fire victims.

By 2:00 P.M., the cashiers had processed eighty-nine people out. New mothers and their babies were sent home. Families were asked to pick up members from the psychiatric ward so beds in that ward could be used by fire victims.

Head nurse Kathy Brown recalled an incident in which she went into the hospital's obstetrics ward and found "some guy sitting in the birthing bed, just like a king, surrounded by cribs and lights. He wound up with one of the better beds in the hospital. I don't know how the people felt who were put in the psych ward. I guess a bed's a bed."

The less serious cases were taken back to the Las Vegas Convention Center.

"So many people stopped us that day and said thank you for taking the time to be so patient," said Southern Nevada Memorial Hospital charge nurse April Row. "After working like a Trojan on my day off, that was really nice."

Row said that she would never forget one man from Leeds, England. He told her that he'd been to Las Vegas before but that this was the first time he'd met "real" people.

He said, "I'm not glad it happened, but at least I've seen the other side of Las Vegas. These clothes I have on my body I'll cherish and take home with me. They were given to me when I left the hotel without so much as a blanket."

She will also remember the man from Pittsburgh who had the presence of mind to dress and put his medicine in a plastic bag while he waited for the firefighters to take him from his room. Through his labor union, he'd been trained in disaster procedures.

"We really tried to treat these people and their relatives with compassion and kindness," she said. "It's traumatic, the carbon monoxide, the things that happened to them, and being away from home. And most of them were asleep when it happened."

Southern Nevada Memorial handled 101 patients that day.

At Valley Hospital, the scene was much the same. Patients coming from the Convention Center emergency base station clutched pieces of yellow ruled paper on which their names and symptoms were scribbled.

Valley ran out of beds by 2:45 Friday afternoon and was forced to refuse more patients. More than a hundred were treated in the emergency room and returned to the Convention Center.

"Every bed is full," said Valley community-relations director Vicki Bertollino. "We're overloaded. We even had to request four respiratory therapists from Panorama City Community Hospital in California."

By midafternoon, tired doctors, nurses, and paramedics found a moment for sandwiches and coffee. Transportation pools were set up to take those who'd been treated back to the Convention Center and from there to hotels, motels, and private homes.

Some, nurses said, refused to go anywhere but the airport.

Dr. John Batdorf said, "I venture to say that if this had happened in any other city in the world, it would have been the disaster of the century. We were prepared."

At the Convention Center, volunteers from the suicide-prevention center and other twenty-four-hour hotline workers attended banks of telephones.

Centel (Central Telephone-Nevada) had installed a fourteen-line Red Cross switchboard so workers could find rooms, medical treatment, and outgoing flights for fire victims. A dozen telephones were set up for local calls only, and twenty-four operator-direct lines were hooked up for long-distance calls, eliminating the need for coins.

In the course of that long day and night, Centel would process 912,071 long-distance calls, breaking its record for one day's traffic. Operators, many of whom returned to work on hearing of the fire, handled 122,447 calls, another record.

"The equipment held up better than on a normal day," said Sal Cinquegrani, Centel representative. "The computer, if we want to humanize it, was probably saying to itself: 'Hey, we gotta get this job done!' "

Aiding the relief agencies, offers of help came from every category of business. Taxi companies provided free transportation and child-care centers provided free baby-sitting so nurses and volunteer workers could help with the victims. Catering-company owner Chris Karamanos sent trucks to deliver food. Motels offered free rooms, recreational-vehicle dealers lent motor homes, and an optical company sent word that it would make new glasses and contact lenses for people who had lost theirs in the fire.

United Way agencies, the Red Cross, and the Salvation Army distributed a flood of supplies from hotels, restaurants, and stores: more than 7,000 sandwiches, 852 gallons of coffee, and thousands of gallons of fruit juice, milk, and soft drinks.

And the Salvation Army issued sixty-six hundred items of emergency clothing to fire survivors. As its original supplies were depleted, people brought in more. By the end of the day, the Salvation Army had half again as much clothing as it had started with.

As the day wore into evening, weary doctors, nurses, and volunteers were replaced by an incoming shift.

At Desert Springs Hospital, Dr. William Berliner, frantic all day about the fate of his daughter, found a note under the windshield wiper of his car. She was all right.

John N. Murdoch was given the job of finding refrigerated trucks to serve as temporary repositories for bodies. After frantic efforts to locate some, he rounded up from Double G. Enterprises, Garrett Trucking, Desert GMC, and Anderson Dairy, enough big semis to serve the need.

As the grim search for bodies continued in the stricken hotel, veteran *Sun* reporter Harold Hyman described their removal from the roof, where most of the bodies from upper floors were taken.

"The helicopters flew for three hours in a nonstop, looping, counterclockwise circle; up from the parking lot to the roof, empty, and back down again with three bodies on stretchers. . . . As the helicopters approached, their prop wash forced coroner's deputies to stand on the empty body bags to keep them from blowing away. . . . Several women were in nightgowns; men were also in various states of dress. There were absolutely no children."

At the morgue, preliminary identification was attempted. One woman, Luz Patino, of Irvine, California, was erroneously listed among the dead because her convention name badge was found near a fallen guest.

Patino and her husband, Rafael, had been rescued from the sixteenth floor and were safe at the Las Vegas Hilton.

For other families waiting to learn, the fate of their loved ones the outcome was tragic.

Helicopters and trucks delivered the bodies to the Clark County Morgue. Insurance man Bill Dietz, who had listened all day to reports of the fire, was struck by the reality of it that afternoon. Returning to his office, he looked up to see a big air-force helicopter coming down for a landing. He thought, *Now, what's that doing in this neighborhood?* Then he remembered that the morgue was nearby, and he realized that the burdens workers were carefully unloading must certainly be bodies from the hotel.

Fire Captain Smith and his exhausted men left the scene of the fire at six o'clock and headed across the street, back to Station 11.

When they arrived, they discovered that thieves had ransacked their food stocks and a firefighter's locker. The crew's chow fund had been taken out of an old coffee can.

At 6:30 P.M. Friday, *Sun* reporter Mary Manning was at the morgue.

"The door opened on a nightmare. At least sixty gurneys stretched across the room, in the halls, even in the offices.

"As I slipped past those lifeless bodies, one woman caught my eye and held it. She looked so peaceful, as if she were sleeping. I kept waiting for her to wave the stiff hand hanging in the air, to smile at me, to open her eyes and sit up.

"She never moved.

"A doctor standing nearby surveyed the scene. He was white faced and still and didn't look at me until I took his hands in mine.

" 'How are you?' I whispered, as we held on to one another. The doctor shook his head in disbelief. Then he began to cry."

All over the city, people were grieving for the fire victims. On Friday evening John and Renée Jones, still shaky from the afternoon spent on the body detail and still covered with dirt and soot, went to Temple Beth Sholom to pray.

Rabbi Kalman Appel recited the Kaddish, and the congregation joined him. Then there was silence.

11 . . .
Questions from the Press

During the long, sad night, as people seeking to learn the fate of missing relatives and friends continued to arrive in Las Vegas, doctors and morticians ushered family members through the morgue.

After several inquiries throughout Friday and Friday evening, Mexican President José López Portillo was informed that among the identified dead were his friends Fernando and Susanna Lobo Morales. He dispatched his private plane, and at 2:30 A.M., in the cold darkness of early Saturday morning, the jet arrived at McCarran International Airport to take their bodies home to Mexico.

The couple had come to the MGM Grand, a favorite holiday spot for many Mexicans, in bereavement: Only four months earlier, Fernando Lobo Morales's father and the couple's young son had perished in a jet crash.

The youthful industrialist had been president of Protexe, a huge Mexican manufacturer of oil-drilling platforms.

In all, thirteen Mexican nationals died in the disaster. In its aftermath, there would be angry criticism in the United States and Mexico over the absence of any escort of Nevada officials as the bodies of the Lobo Moraleses and other Mexican guests were transported to their homeland.

Another irony of the tragedy was that shortly after dawn on Saturday morning, Barbary Coast employees telephoned Metro police and reported that a man wearing a soot-stained MGM Grand security-guard uniform was playing their slot machines with blackened coins.

Police arrested John R. McManus, a former New York Police Department sergeant, who was alleged to have been on duty guarding the fire-ravished casino.

Found in his possession were $346 in tarnished MGM Grand dollar tokens, $652 in burned and dampened currency, and a half dollar.

Saturday morning was cool and brisk. The sky was clear and the sun was shining. A brisk breeze was blowing. On the sidewalk outside the hotel, discolored slivers of metal and shards of glass danced about. Draperies fluttered through the resort's broken window panes; yellow bed-sheets, still suspended from balconies, were a mute reminder of the day before.

Barry and Trudy Richards shivered in the deserted Strip resort's shadow. It was shortly after nine o'clock, and they were jockeying for position in the line that was to take them up twenty-three flights of stairs to their room. They had come to retrieve the clothing and valuables left behind in the emergency.

The California couple had checked into the Dunes Hotel on Friday after leaving the Sunrise Hospital emergency room. When registering, they had asked for a room "no higher than the second floor." The desk clerk was understanding, and when they got to their room, Barry looked out the window. It was just a nice, easy jump to the ground.

A large number of guests, some still in the blackened clothing they had worn since escaping from the hotel the day before, had come earlier than nine, hoping to reclaim their belongings and leave town. But after two hours, the line wasn't moving.

Steve Pelzer, assistant hotel manager, spoke through a bullhorn, pleading for patience. "It will be a very slow process, but we are going to start going up in a few minutes. Move back. Just give us some room here, please."

Off to the side, MGM Grand employees, among them pit bosses, floor men, and dealers, stood ready to escort people to their rooms. They were dressed in T-shirts, blue jeans, and sneakers. When this weekend was over, they'd be standing in the unemployment line.

The employees were there as a security measure, to prevent looting and to help bring luggage down the stairs. But fire officials had not yet given the approval for anyone to enter the building.

Pelzer grabbed the bullhorn again. "Ladies and gentlemen, we're getting very close. In about twenty minutes, we're going into the building. Our hands are tied until the building is released to us. Just sit tight for a few more minutes."

Pelzer was greeted with jeers and catcalls, but the people waited—they had no choice.

At 11:30 A.M., the guests, who had by now formed a sizable and diffuse human mass in front of the MGM Grand's Arcade entrance, were ordered into single-file lines, each line corresponding to a floor of the hotel.

In the crowd, the anxiety and frustration of the last twenty-eight hours was boiling over.

Some of the guests accused guards and employees of dealing in favoritism.

"They were not in line," yelled one outraged person to a hotel employee, pointing at a couple at the front of the line. "You made friends with them and let them go in. We've been here since 7:30 this morning."

It didn't help. They had to wait.

Barry and Trudy Richards's line eventually moved forward, and the couple trudged up the stairs to the twenty-third floor. The stairwell still smelled of smoke. A grayish film covered the walls. Soot stains radiated from under doors.

After the tiring hike up the stairs, the Richardses made it to their room. Though smelling of smoke, their luggage and belongings were intact.

Others were not so lucky. A large number of MGM Grand guests looking for possessions in their smoke-ravaged rooms reported missing wallets, handbags, and jewelry to law-enforcement officers at the scene.

"They stole my rings, my camera, and my watch," said Henry Greenspoon of Montreal, sitting on his retrieved suitcases. "What a mess."

There were so many complaints of thefts that Metro

police sent an officer to the hotel to take reports. Officer Alan Fisch worked all day Saturday and Sunday, from a desk near the jai-alai fronton. Police arrested one looter found with sixty-six hundred dollars in cash and jewelry from abandoned rooms.

As late afternoon closed in toward evening, people were pressing against one another for warmth, waiting for lines to inch forward.

Metro-police officer Dan Harness, walking across the parking lot, was visibly distressed. "There's some stealing going on up there," he fumed. "There's one lady who lost a thirty-thousand-dollar ring, and we don't know who got it. We won't know how much was lost until later, when people go home and look in their suitcases and really see what's missing.

"But that's not what I'm worried about right now. I'm going to see if we can get something from the Red Cross. It's getting cold, the people are getting hungry, and the older folks, their legs are locking up from standing so long."

Frustration mounted as the sun came close to setting behind the mountains.

"We've been here all afternoon," complained Cindy Vander Maten of Fort Dodge, Iowa. She and her husband had neither money nor clothing. "People kept budging in front of us in line and giving their sad story. Now we're being sent away until tomorrow. I'm just mad."

Neil Johnson from Ann Arbor, Michigan, waited in line for six hours and was turned away at about 4:30 that afternoon.

"I've got a chartered flight out tomorrow morning, and it can't be canceled," he said. "So we're going to end up leaving without our luggage. I've got hundreds of dollars up there I'll never see again."

On the other hand, Rich Chappel, executive director of the California Tow Truck Association, based in Los Angeles, appeared to take the grueling wait in stride. "I'm not bent out of shape. What else can you do? If nothing happens in the next half hour, I'm going to go watch the football game."

County Fire Chief Roy Parrish and Mike Patterson, who heads the investigative division of the Clark County Fire Department, called a five-o'clock press conference Saturday afternoon. The reporters were to meet in the Mark Twain Room at the Barbary Coast to be given an update on the fire. But no one really knew what would be said.

Many of the reporters were developing a hypothesis that a grease fire in the Deli kitchen had sparked the blaze. Others, who had heard reports that a keno board in the hotel exploded before the casino became engulfed in flames, pointed at an electrical failure.

"There are a lot of questions we won't have answers to for many weeks or months," Parrish told reporters at the meeting. "But we have established certain facts, which we will relate to you."

The fire, Parrish said, started in the attic above the Deli and was "electrical in nature."

He told reporters that the fire had smoldered for several hours, generating heat in the open area above the casino-level restaurant and its kitchen.

When the fire broke through the kitchen ceiling, exposing the combustion process to the air, the sudden rush of oxygen fueled the glowing embers and generated a giant backdraft that sent a wall of flame rushing along the casino ceiling.

The mass of fire traveled the casino twice. First it roared to the front of the building, leaping down into the hotel lobby after the ceiling near the main entrance collapsed. Then it made the return trip, ripping through the 140-yard long casino like a giant torch. The enormous room turned into a searing oven.

"It will take us a week to find out why the fire spread so fast," Patterson said. "The problem with this fire is the time it had to smolder. The fire wasn't discovered until it broke through the ceiling in the Deli area."

The twelve-foot gap between the casino ceiling and the subfloor teems with air-conditioning ducts, conduits, pipes, and insulated cables. Tucked into their intricate webbing of metal and plastic was some of the electrical wiring that fed the casino lighting, the keno boards, and the slot machines.

As the scorching wall of fire sped along, it set pipes and ducts ablaze. Lights in some areas of the building went out as wires above the ceiling melted. Seconds later, the ceiling over the front entrance collapsed, and flames fell into the casino.

The wall of flame then blocked the front of the casino. At the other end of the casino, another raging fire had already forced Captain Rex Smith's team of firefighters out the side entrance.

The front entrance, with its array of open glass doors, provided an enormous source of oxygen. This fed the wall of flame and gave it a Herculean shove that sent it through the casino in the blink of an eye.

Patterson told the press, "Based on our preliminary investigation, which included witness statements from people who saw the fire break through the ceiling, we find that this is the way it happened. We have experts from the National Fire Protection Association who will examine the scene and determine how the fire spread and why it propagated so quickly."

The press was quick with a barrage of queries:

Q: How did the smoke get to the high-rise?

A: There were three elevators below the casino floor. The doors were open, and the touch mechanism was melted by the heat. There was a natural draft up the elevator shaft. When the smoke reached the top, it mushroomed down— [filling] each floor from the top down.

Q: Was MGM Grand management ever advised to update its fire-protection system?

A: The MGM was not notified of the changes in the codes. I strongly believe that a grandfather clause should be written into the code, but the judges will throw it out.

Q: What is the latest death toll?

A: Eighty-three.* I don't know exactly how to put this without it sounding wrong, but there's a positive aspect to this. We had close to eight thousand people in that hotel. With eighty-three killed, we had a one-percent loss of life.

When I got there, I thought we were going to lose hundreds of people. So in a way, we were very fortunate.

Q: Could more lives have been saved?

* The eighty-fourth body, that of María Lucy Capetillo, was discovered Saturday night.

A: If more people had stayed in their rooms, [fewer] people would have died.

Q: Was the fire at the MGM Grand a freak accident?

A: This particular fire can happen anyplace in the United States. Las Vegas is not unique in its problem.

By Saturday night, crowds had thinned at the Convention Center, but civil-defense and volunteer workers still labored over telephone banks and hotel census lists as calls continued to pour in from all parts of the world. Of the evacuated guests, thousands had started for home, many without their possessions. Hundreds of fire victims were in hospitals, and hundreds more were sheltered in hotels, motels, and private homes while waiting for the return of luggage or for the recovery of friends and relatives too ill to travel.

Convention Center disaster workers that grim weekend were also trying to determine the fate of about three hundred people still unaccounted for—people who were believed to have been in the hotel when the fire started but who had not yet turned up on survivor lists or on hospital or morgue records.

So it was that Randy and Brad Keller flew in from Indianapolis to seek word of their parents and duly arrived at the Convention Center.

The brothers went first to the hotel and after some argument, were admitted to their parents' room. Neatly packed suitcases were all that they found. Their apprehension increasing, they next went to the morgue. Astonishingly it was closed, and a notice posted said that it would reopen at eight o'clock Sunday morning.

Overwhelmed with grief and rage, still hoping their parents were somehow alive perhaps "wandering somewhere in a daze," the Keller brothers and some forty other people denied access to the morgue waited at the disaster center while volunteers Frank Weinman and Rusty Feuer called state legislators and county officials. Their pleas had no effect until they reached County Commissioner Thalia Dondero, who cut through the red tape. The morgue was reopened at 11:00 P.M.

Then, for the Kellers, the last bit of hope vanished. Their father and mother, Jac and Blanche Keller, had apparently been breakfasting in the coffee shop when the holocaust began. Their bodies, burned beyond recognition, were identified from dental charts and items of jewelry.

MGM Grand corporation officials, who had avoided media queries and were, in a sense, "unavailable for comment" from the moment the casino was filled with flames, eventually met with reporters during an emotionally charged news conference Sunday morning.

Corporate chairman Fred Benninger appeared at the meeting along with Alvin Benedict, president of MGM Grand Hotels Inc., and Bernard Rothkopf, president of the Las Vegas MGM Grand. Benninger, acting as spokesperson, told reporters that the lack of a comprehensive alarm system at the hotel "may have been a blessing in disguise." He argued that warning devices would have sent terrified guests streaming into smoke-filled hallways, where many might have perished.

He said that the hotel's fire equipment was within code

and that the corporation deemed further safeguards unnecessary.

Codes, he said, would again be met when the hotel was rebuilt, but no decision had been made on whether to order extra fire equipment. He set the target date for reopening the hotel, as well as the 782-room addition, at July 1, 1981.

Benninger adamantly denied rumors that MGM Grand executives had been advised by the Clark County Fire Department to upgrade sprinkler and alarm systems beyond minimum code standards.

At Saturday afternoon's news conference in the Barbary Coast, Fire Chief Roy Parrish had fallen silent when reporters asked him if he had ever met with MGM Grand executives to suggest such improvements.

Parrish had hinted to some reporters that even if he had made recommendations, he would have been in no legal position to demand that the MGM Grand carry them out.

"As long as the hotel is up to code, my hands are tied. I can make all the recommendations I want, but they don't have to do a thing."

The noncentralized alarm system, Benninger said, never sounded because "part of the alarm burned off. If the alarm had gone off, guests would have gone into the hallway," where they would have been met by an opaque cloud of smoke and died.

However, many firefighters, interviewed after Benninger's comments hit the presses, said the corporate executives were "just trying to cover their asses by saying that."

According to one ranking county fire official, "If the alarm had gone off early on, when it was supposed to, people would have been in the halls, down the fire stairs, and out of the building before the smoke got so thick that no one could see."

At one point in the news conference, Benninger threatened to walk out. As he was getting up to leave, a man with an NBC television camera cradled on his shoulder yelled: "Eighty-three people died in your hotel, and you're going to walk out on the press?"

Benninger sank back in his chair, and the meeting resumed.

"We certainly don't know why the fire spread so fast," Benninger said.

He said that an employee in the catwalk above the casino had not smelled smoke when he was attending the eye in the sky thirty minutes before the gambling hall was overrun by flames. This account was a confirmation, or at least a reiteration, of something county fire inspector Mike Patterson had said the afternoon before.

Benninger said that all the exits were clearly marked and that employees had been told what to do in the event of a fire. However, there had been no formal fire drills.

The harried hotel executives did not say how much money was recovered after the blaze.

"Five hotel staff died in the fire," Benninger said. (The number of MGM Grand employees who died in the hotel was subsequently found to be nine.) "None of them died trying to save the money, although the casino staff was under instructions to secure the money" in the event of a disaster.

*   *   *

About thirteen hundred jobless MGM Grand Hotel employees lined up outside the Nevada Employment Security Department in downtown Las Vegas on Monday, while hundreds of others flocked to the Culinary Union hall seeking new jobs.

It was the largest single day of claims handled by the state of Nevada, surpassing the total number of Aladdin Hotel workers who had sought relief earlier in the year when that Strip resort's casino was closed by court order.

A dozen employment-security officials from northern Nevada were flown to Las Vegas to help the local office process the claims, which would be filed over a three-day period.

Many of those applying for unemployment compensation were blackjack dealers, cocktail waitresses, and bell captains, whose low salaries are offset by tips, but they would receive compensation based only on their salaries.

Out-of-work MGM Grand employees waiting in line expressed shock at the sudden loss of their jobs and uncertainty about future work prospects.

Chris Gangemi, a waitress and mother of two, said, "I told my kids we're going to have to cancel Christmas this year. I don't know what I'm going to do now."

Others were more optimistic. Mark Gordon, a singer in what was to have been the hotel's new *Jubilee* stage spectacular, remarked philosophically, "At least I'm alive. The people in the hotel that died in the fire can't draw unemployment."

Messages of sympathy were dispatched by President Jimmy Carter and President-elect Ronald Reagan, a message from Pope John Paul II was conveyed in a telegram to Nevada Bishop Norman F. McFarland from John Cardinal Cassaroli, papal secretary of state.

At the Guardian Angel Cathedral, situated off the Strip about a block north of the Desert Inn, hundreds of people gathered for a noontime memorial service for the eighty-four dead.

Bishop McFarland delivered a short, moving homily to a standing-room-only congregation at the mass. The bishop and twenty priests celebrated the Eucharist in remembrance of those who had died in the second worst hotel fire in this country's history. (The worst U.S. hotel fire occurred on December 7, 1946, when the Winecoff Hotel in Atlanta burned and 119 people died.)

That night, the wreckage of the MGM Grand sat dark on the brightly lit Las Vegas Strip.

Every once in a while, a whitish glow would materialize behind a shattered windowpane as a security guard poked his flashlight into a room.

A harsh, cold wind whipped across the desert late that Monday night. The fifty-mile-an-hour gale blew curtains out of windows and sent pieces of glass hurtling down onto Flamingo Road.

Three days after the tragic fire, the deserted MGM Grand still posed a hazard, and police closed the street.

12 . . .

Code Violations

The final death toll was eighty-four. According to the Clark County Coroner's report, one victim fell to her death, eight died in the fire, and the other seventy-five people died of smoke inhalation from breathing the poisonous gases released, in part, from burning plastics. At least half of the hotel's furnishings, fittings, and decorations were made of plastics.

Smoke inhalation can be fatal because life-sustaining oxygen in the bloodstream is replaced by carbon monoxide, which prevents the central nervous system from functioning.

Some of the MGM Grand guests were asleep or barely out of bed when they were overcome by the toxic smoke and gases. Others, those in the elevators or running through the halls, would have first become dizzy and lightheaded. Depending upon the concentration of the smoke particles and on the victims' physical condition, dizziness would sooner or later have been supplanted by confusion and irrationality, then by unconsciousness. Some would have expired in a few minutes; for others, comatose, death could have come after an hour or more.

Before the ashes were cold, as searchers still combed the hotel looking for bodies, the largest question of all was beginning to emerge. Why had eighty-four people died? How could the tragedy have been averted? Why had the hotel, only seven years old and one of the most lavish resorts in the world, become a deathtrap?

On November 21 and during the days that followed, officials from fire departments throughout the United States, Canada, and Mexico came to Las Vegas to study the tragedy and learn from it.

Firefighters from Mesa, Arizona, were among those who observed the efforts of local firefighters as the disaster was happening. Many of those attending a national fire marshals' conference in San Diego that week came to Las Vegas to get a first-hand look at the fire scene.

Some traveled even farther. Nils Fröman, an assistant chief of the Statens Brandnamnd, Sweden's board of rescue and fire services, left Stockholm on November 25 to study the MGM Grand fire.

"If you think of the circumstances, they were very lucky," Fröman said. "It could have been worse than it was."

Fröman and a doctor toured the fire scene and spoke with many of the officials who had had a part in fighting the blaze and treating the victims.

"We take experience from it. We look at the medical

side and the accident itself. What we learn we take back to Sweden, where we will apply what we've learned to the regulations we already have."

Fire prevention is a comparatively recent science. The bibles of the craft are the national fire and building codes. When inspectors make their annual rounds at structures like the MGM Grand, they use the code books as their guide. After the tragedy the inspectors' initial findings concluded that the building had been built within codes—and relatively recent codes at that.

A showpiece of modern architecture and construction methods, the MGM Grand was supposed to offer a measure of protection to occupants in the event of fire. Only later did evidence emerge pointing to numerous code violations.

The most glaring deficiency in the smoldering casino was fire sprinklers. The MGM Grand was constructed in 1973 under 1970 building codes that did not require sprinklers in buildings of its type. Chief Parrish pointed out that if the MGM Grand had been fully outfitted with fire sprinklers, the only damage wrought by the small electrical fire would have been a puddle of water.

The hotel did not totally lack sprinklers. The basement and shops area, the twenty-sixth-floor banquet rooms, and parts of the main floor—but not the main casino—were sprinklered. Where flames approached sprinklered areas, they were stopped by the shower of water. In one area, heat from the fire was so intense—estimated at more than three thousand degrees Fahrenheit—that sprinklers were activated 120 feet from the closest flames.

The MGM Grand is only one of several major Las Vegas hotels that were virtually without sprinklers in November of 1980. A report prepared for the Clark County manager's office named eleven other hotels as lacking sprinklers in their casinos. These included such famous resorts as the Las Vegas Hilton, the Flamingo Hilton, the Desert Inn, and the Riviera.

County Manager Bruce Spaulding ordered a law drafted that, if passed, would require older hotels to be retrofitted with sprinklers. The county Board of Commissioners would have to approve the proposed ordinance. As the rules stood in 1980, all hotels beginning construction or submitting building plans for approval in 1973 or thereafter were required by building codes to install sprinklers. Construction on the MGM Grand began in 1972.

Asked why a retrofit ordinance had not been considered before the fire, Spaulding replied, "There was no chance in hell it would have been successful. Retrofitting is very expensive. It's hard to say, but the fire gives us an opportunity to get done some of the things we wanted for a long time."

County commissioners immediately began to consider retrofitting. "You've got to determine two things," said Commission Vice-Chairman David Canter. "Is it physically possible to retroactively place the systems, and how much does it cost?

Canter readily admitted the safety benefits of retrofitting; however, he stopped well short of advocating the politically risky ordinance. "Obviously," he said, "the 1980 car I have is built safer than the 1979 model I bought for my wife, and *it* is safer than my 1971 Ford. Each year, the cars get safer, but I'm not willing that the Ford Motor Company should go back and put these safety devices in it."

State Fire Marshal Tom Huddleston predicted a legal challenge by the hotel industry if laws were adopted ordering retrofitting. However, it appeared that the protests might be muffled as resorts tried to bounce back with a strong public-safety image in the wake of the MGM Grand fire.

Huddleston estimated the cost of installing sprinklers in the new buildings as being roughly equal to the cost of carpeting. For example, the Hacienda Hotel, at the south end of the Las Vegas Strip, installed a comprehensive fire-control system at a cost of more than $250,000. The system includes sprinklers, alarms, smoke detectors, and public-address speakers in each of the resort's 265 rooms.

Ironically, the small Barbary Coast Hotel adjacent to the MGM Grand had a small electrical fire the same morning as the MGM holocaust. Owner Michael Gaughan said the blaze was quickly extinguished. The Barbary Coast, opened in 1979, was required by code to have comprehensive safety features.

Former Clark County Fire Marshal Carl Lowe, who oversaw the MGM Grand's early construction phase, said that he had urged owners of the resort to install a comprehensive sprinkler system during construction, but they replied that it would cost too much money. "All they wanted to know is what the code stated," Lowe commented. "A fireman can see a lot more than the average citizen. But with builders, all they see is dollar signs. The building code was a little outdated for that time in Las Vegas. I would have loved to see sprinklers installed in the MGM."

The MGM Grand's new 782-room addition must meet the 1979 fire and building codes, which call for complete sprinklering and other fire-safety systems. Fire inspectors, reading the building code even more closely, said that they interpreted the code not only as applying to the addition at the time of its construction, but as requiring retrofitting of the existing structure was well. Architects of the project disagreed and reportedly wrote a letter to Clark County Fire Marshal James Barrett in February 1980, asking him to acknowledge "that there is no legal requirement to separate or add fire sprinklers to the existing facilities."

The 1970 code stipulates that automatic sprinkler systems are required on display and exhibition floors in excess of twelve thousand square feet. Although the MGM Grand casino is much larger, a specialist for the International Conference of Building Officials (ICBO), the organization that drafts the codes, indicated the requirement had never been meant to apply to casinos.

"It wasn't a specific requirement," said ICBO Technical Director Don Watson. "The code didn't apply to a casino floor. It wasn't intended to mean that." The clause was intended to refer to trade shows and conventions that use large exhibition rooms. The ICBO drew up the clause after a fire tore through an exhibition hall in Chicago decorated with paper streamers that burst into flames.

Fire officials agreed that flammable contents, and not the structure itself, led to the quick spread of flames throughout the MGM Grand. The fire department sent samples of carpets, drapes, and other furnishings to the United States Bureau of Standards for analysis. "If there is any material in there that didn't meet code, we want to know about it," declared Deputy Fire Chief John Pappageorge. "If it did meet the code, then maybe we should look into changing the code itself." When something burns, heat causes the substance to give off gases. It is these gases, not the substance itself, that burn. Extreme heat, such as that experienced in the spread of the MGM Grand blaze, causes gases to be emitted. No matter how well the materials are flame-proofed, the fumes will often burn.

Interior furnishings and lack of sprinklers were not the only major reasons for the spread of the fire. Investigators found evidence of violations of the fire and building codes. Parrish and County Manager Spaulding both noted that holes were found in walls designed to contain fire. Holes were cut to accommodate the catwalk above the casino ceiling designed as a platform for the eye in the sky.

Parrish said of this, "These violations are always serious. People should never cut holes in fire walls for any purpose whatsoever." He later added, "I'm sure smoke got in through the areas that were cut into the dry wall."

But the *Las Vegas Sun* said in a copyrighted story that violations reported by inspectors went far beyond mere holes in walls. The newspaper reported that inspectors had found the hotel's fire-alarm system rigged so that it would not sound for five minutes while security guards checked out any report of a fire. In addition, the hotel's main air-circulation equipment was bolted in such a way as to prevent proper operation of smoke dampers that could have potentially cut down on the circulation of smoke.

One of the major investigation teams represented the National Fire Protection Association, which writes fire codes. They were to be one of the groups to prepare a complete report on their findings.

Investigators said they believed that some of the purported violations in the MGM Grand could conceivably extend back to construction. After studying records dating back to that time, the *San Francisco Examiner* reported that an agency which advises insurance companies criticized allegedly improper coating of steel supports. Herman Speath, supervisor of codes and standards for the Insurance Services Office of San Francisco, alleged the coating might flake off in the event of fire, leaving the steel unprotected. Without this protection, steel would buckle under one-thousand-degree-Fahrenheit temperatures within five minutes. Speath's finding evolved from a 1973 visit to Las Vegas during which he decided to stop off at the MGM Grand construction site. During the MGM Grand's construction, the county fire department had some of its own disagreements with the county building department. County Fire Marshal James Barrett wrote to the building department about the discovery of a showroom aisle at the MGM Grand that was less than three feet wide "in direct violation of . . . the Uniform Building Code," which, he said, the building-department chief at the time, John Pisciotta, was supposed to enforce.

Contacted by the *Las Vegas Sun,* Pisciotta, the owner of a Las Vegas construction company, defended his record on the MGM Grand. Pisciotta cautioned against second guessing what safety devices should have been included in the MGM Grand beyond code requirements. "You can make suggestions all day long. You can't enforce anybody's suggestion unless it's the code. It's that simple. It's just like a cop. You book somebody who is arrested. You don't make up the rules.

"[The MGM] has a lot less problems than other buildings," he went on. "It was probably one of the best jobs I ever saw when I was building director. I've got nothing to be afraid of. I've got nothing to hide. There was no problem at the MGM. Code requirements were met. I was the enforcement officer, and I don't enforce suggestions."

(Two months after the fire, Pisciotta was named as one of several defendants in a two-billion-dollar damage suit.)

Asked whether other Las Vegas resorts constructed under old codes should be ordered to install sprinklers and fire alarms, Pisciotta replied, "After the *Titanic* sank, did they stop building ships?"

No, they did not stop building ships, but they did make them safer.

When the *Titanic* started its maiden voyage, it was the world's finest and newest luxury liner. So safe was it considered, that it was thought to be unsinkable, and it carried enough lifeboats for only a fraction of its passengers and crew.

It was, however, up to the code of its day. The regulations covering the number of lifeboats had been written in the nineteenth century, long before the era of superliners.

Other factors contributed to the disaster: Intent on breaking a speed record, the captain ordered full speed ahead, and the doomed vessel was racing through known iceberg zones.

And when it struck the fatal iceberg and began to sink, a ship only ten miles away stood silently by. The vessel's wireless operator had gone to bed, leaving his radio unattended.

But they were all within the letter of the law—the *Titanic*'s builders, its captain, and the ship so close that it could have rescued all aboard, had it maintained a twenty-four-hour radio watch.

In the uproar that followed the *Titanic* tragedy, the sea laws were rewritten.

How good are hotel fire-safety laws? The record is uneven, experts agree. As many as eighteen thousand state and local code jurisdictions across the United States form a patchwork of regulations.

And then there are the violations of those regulations.

After six hundred hours of study, fire experts determined that the primary cause of the MGM Grand fire was a short circuit in a brief stretch of wire in the wall of a serving station in the Deli restaurant. For the giant hotel, with its almost two thousand miles of wiring, this was its Achilles heel. Heat buildup in the wiring caused a small fire that smoldered for an undetermined number of hours before it burst into view. Then why did it spread so swiftly?

On December 15, 1980, Clark County Manager Bruce Spaulding issued a preliminary report on the tragedy to Robert D. Weber, director of the Clark County Department of Building and Zoning. The report confirmed earlier *Las Vegas Sun* allegations that hundreds of code violations at the MGM Grand "contributed adversely to the life safety of the hotel occupants."

Weber noted in the report that smoke paths leading to each level of the high-rise hotel were created by penetrations in air shafts, open hallway soffits, holes in corridor fire walls, a lack of fire dampers in air shafts, plastic pipes penetrating fire walls, and insufficiently vented elevator shafts.

On January 16, 1981, the National Fire Protection Association (NFPA) reported that special structural features designed to absorb the shock of earthquakes may have assisted the spread of poisonous gases and smoke through the twenty-six-story high-rise.

Weber, referring to the NFPA's preliminary report, said that the seismic joint shafts, built to enable a high-rise to withstand the shock of ground movement, were apparently built in violation of county building codes of the early 1970s. He contended that the faulty shafts conducted smoke up the building in much the same way as open elevator shafts and stairwells did.

"We're continuing to study the overall interaction of portions of the building. We've identified a number of penetrations that were abnormal," he said.

A spokesman for the MGM Grand, echoing a familiar tune, would say only, "We were built to code."

Other contributing factors: open elevator doors that led to excessive smoke entering corridors; the extensive use of improperly treated wood throughout the structure; improperly designed kitchen exhaust-duct systems; inadequately protected steel; air-conditioning and ventilation systems that did not automatically shut down under intense heat levels; and ceilings and attic areas built with materials that would hasten the spread of flames. On the credit side, officials noted, were the automatic sprinklers installed in certain corridors, showrooms, restaurants, and kitchens adjacent to the fire, which prevented its spread within ten to twenty feet of the first sprinkler in contact. The December 15 report said, "The sprinklers performed in an excellent manner."

As charges and counter-charges flew, stock-market analysts began trying to look months and even years into the future of the company. The experts who chart the fortunes of public companies such as MGM Grand Hotels, Inc., agreed that the MGM Grand's financial future would depend largely upon the size of the corporation's insurance package.

A little more than a month after the fire, thirty-six lawsuits had been filed against MGM, in several federal jurisdictions. Damage claims ran into hundreds of millions of dollars. Furthermore, dozens of additional suits were filed in various state courts around the country.

Attorneys connected with the cases agreed that there could eventually be as many as four thousand plaintiffs. Litigation of the suits was expected to take at least four years. The noted San Francisco attorney Melvin Belli predicted that damage awards would eventually amount to more than one billion dollars.

In the weeks following the fire, United States District Judge Roger Foley found himself dealing with a variety of motions. He rejected a suggestion that he tie up the resort's two hundred million dollars in assets while the litigation was in process.

During a hearing in December 1980, Foley gave lawyers permission to dismantle an entire twenty-third-floor MGM guest room, including the walls, ceiling, and carpeting. The attorneys said they wanted to use the room where four persons died as evidence. In the first days following the disaster, squads of attorneys met for breakfast at the Dunes Hotel, then crossed the street to the closed-off MGM Grand, where they donned hard hats and searched for evidence in the structure. A photographer accompanied them on their rounds, documenting the scenes of death and injury.

Among suits filed less than a month after the blaze was one by Columbia Pictures Inc. that charged the MGM was unsafe and underinsured before the fire. The suit attempted, unsuccessfully, to delay the MGM's stockholders' meeting and was intended as a roadblock in the path of Kirk Kerkorian in his attempt to enlarge his twenty-four-percent stake in Columbia.

A class-action suit was filed by Los Angeles lawyer Fred Kunmetz, in which he asked damages totaling $250 million on behalf of 350 Mexicans who had been staying in the MGM at the time of the disaster.

Two months after the fire, in the wake of dozens of other actions, Los Angeles attorney Harold Sullivan, representing 255 plaintiffs, filed suit in Los Angeles Superior Court for two billion dollars, charging that his clients were still suffering the effects of toxic poisoning. Named as defendants were MGM Grand Hotels, Inc.; Martin Stern, Jr., the architect; Taylor Construction Co.; its contractor, California Electric Co.; Clark County; and John Pisciotta, the county's former building-department supervisor.

Among Sullivan's clients was Eduardo Hernández, forty-one, of Mexico. Hernández died in a Houston, Texas, hospital, weeks after the fire. The cause of his death was pulmonary edema and the aftereffects of inhaling toxic fumes. According to Sullivan, Hernández was the MGM Grand's eighty-fifth fire victim.

Lawyers and insurance experts predicted that it was scarcely likely that any court would award damages of the record amounts asked for. They further pointed out that if the company went under, no one would get anything. However, attorneys, as well as resort-industry executives who assessed the potential consequences of the blaze, seemed to agree generally with the opinion of a former Nevada gaming controller who said, "This fire is a lawyer's dream."

He was referring to the barrage of complaints that the MGM Grand was guilty of fire-code violations at the time of the blaze. Still, MGM has been consistently recognized by bankers and gaming industry analysts as a strong, conservative company with good management. During the fiscal year that ended August 31, 1980, the company recorded revenue of $307.1 million, with net income of $33.9 million. Both figures were records.

In May 1981, the United States Judicial Panel on Multi-District Legislation in Washington, D.C., ordered all lawsuits across the country stemming from the MGM fire transferred to Las Vegas. Suits in eight federal districts, including Las Vegas, were consolidated under the jurisdiction of the United States District Judge for the Eastern District of Pennsylvania Louis C. Bechtle, of Philadelphia. The panel said the action was necessary to prevent duplication of discovery, to avoid inconsistent pretrial rulings, and to conserve resources of all parties involved in the litigation.

"The District of Nevada, where the disaster occurred, clearly is the preferable transferee district," said a panel spokesperson. Attorneys organizing for the massive legal action formed an eleven-member Plaintiffs' Legal Committee to spearhead the actions. Dozens of other suits remained to be heard in various state courts.

A week after the fire, MGM Chairman Fred Benninger reiterated his own assurance that the company would survive when he declared, "The company is financially sound, notwithstanding the recent tragic events."

Benninger said that the company had $215 million in property and business-interruption insurance and at least $30 million in liability insurance at the time of the fire.

There is little doubt among insurance experts that the $215 million will be more than sufficient to compensate MGM's business losses and repair fire damage.

"In that regard, the company is in very good shape," explained W. O. Slayton, Nevada's chief insurance assistant.

There was considerably more speculation about the amount of liability coverage Benninger said the MGM was carrying.

The *Los Angeles Times* spoke with insurance-company executives and reported, "There is serious question whether the firm has adequate liability insurance to cover its potential losses."

It will be years, however, before these cases are litigated and this can be known for certain.

The *Times* also reported that the experts it interviewed suggested that "the firm may end up digging into its own pockets to pay some liability claims. With average death payments ranging from $250,000 to $350,000, insurance protection of $35 million could well be exhausted by death claims alone. Injury claims would be on top of that."

Claims will, however, be litigated over a matter of years, and analysts familiar with MGM seemed to remain firm in their beliefs that the likes of Fred Benninger and Alvin Benedict would find the means to steer their company through the disaster of November 21.

Ole Johnson, senior vice president of Sayre & Toso, a Los Angeles–based managing general agent that placed a portion of the coverage, said that insurance adjustors who examined the building estimate that claims for property and loss of business revenue will total $85 million.

Johnson called the estimate "a pretty good ballpark guess." Out of the $85 million, losses of business revenue were expected to total $1 million per week for between twenty-eight and thirty-two weeks.

Insurance payments for business interruption are determined by taking gross revenue and subtracting expenses such as payroll and purchases, which are sharply reduced because the hotel is closed.

The hotel's $215 million in combined property and business-interruption insurance is structured this way:

The first $2.5 million is covered by Mutual Fire Marine & Inland Co. The next $7.5 million "layer" of coverage is shared, $5 million by Puritan Insurance Co. and $2.5 million by Pinetop Insurance Co.

Coverage for the next $15 million is provided by Lexington Insurance Co.

The next $25-million layer of property coverage was furnished by Holland American Insurance Co.

Various insurance companies in the Chicago-based Kemper Corp. underwrote the next $125 million in property and business-interruption insurance. The final layer of property coverage was provided by Insurance Co. of North America.

Then in March 1981, MGM Grand Hotels, through a consortium of underwriters, purchased approximately $170 million worth of liability insurance, retroactive to before the fire. The premium was reported as $37.5 million. Retroactive liability insurance of this type is not uncommon in insurance-industry circles. The premium involved would earn money upon investment through the years claims are expected to be in litigation, and this secondary coverage would be tapped only after primary liability insurance coverage was exhausted.

Once the hotel opened again, MGM executives would face other marketing problems as they tried to sell conventions and other large groups on the prospect of returning to the MGM.

MGM stock fluctuated wildly during the days after the fire and there was considerable speculation that financier Kirk Kerkorian, who owns slightly more than fifty percent of the stock, might eventually emerge with ownership of perhaps sixty or seventy percent of the company.

This thinking was based on the belief that Kerkorian was expected to go on a buying spree if stock remained at depressed levels. As it turned out, however, Kerkorian did not buy up extra stock during this period.

New York stock analyst Steve Eisenberg, who specializes in gaming companies, said MGM should eventually recover to its twelve- to thirteen-dollar-a-share level, but he said it would probably underperform the market generally and other gaming stocks until performance indicated that the MGM Grand Hotel had recovered. (MGM Grand did reach a low of 7⅞ at one point, then rallied. As the opening date for the restored hotel approached, the stock was selling for 12¾.)

A fifty-million-dollar restoration-and-repair project was underway as soon as the building was released by county authorities.

Meanwhile, on July 2, 1981, attorney Richard Meyers, associated with the Plaintiffs' Legal Committee, charged that building-code violations had resulted in death and injury on a large scale and should be sufficient to bring criminal charges (in addition to the civil suits) against those responsible.

Clark County District Attorney Robert Miller said that an investigation into possible criminal violations of the building code was proceeding. He explained that sifting through the vast bulk of evdence took time. Consequently, interviews with those involved didn't begin until June 1.

On the following day, Las Vegas United Press International Bureau Chief Myram Borders wrote:

> Evidence that fire and building codes were substandard at the MGM Grand Hotel when a catastrophic fire claimed 85 [*sic*] lives may be presented to the grand jury next month to determine if criminal charges should be filed against those responsible.
>
> Two dozen county employees have been interrogated by the District Attorney's office during recent weeks. The questioning of employees, executives of the gambling resort, and contractors begins this month—on the eve of the July 30 reopening of the 2,076 room hotel currently completing a $50 million renovation and reconstruction project. . . .
>
> Hotel owners recently unveiled a $5 million computerized fire-safety system designed to make the new version of the MGM Grand one of the safest hotels in the world. But county reports

show that prior to the November fire, building and fire codes were substandard.

"It is amazing what some county employees can't remember," said one source close to the investigation. "Some are reluctant to be open because of possible suits . . . some can't even remember whether or not they were inspectors during construction work on the hotel."

13 . . .

Learning from the Tragedy

The Las Vegas holocaust caused newspapers all over the country to call for a long look at their own local fire codes and to write could-it-happen-here? pieces.

The *San Francisco Examiner* wrote that a visitor to San Francisco had better odds of escaping harm. The 1975 life-safety code in that city requires sprinklers in all rooms, smoke detectors, and a fire-department command room. However, the paper concluded that there is still some danger in older buildings not brought up to code.

On November 23, 1980, *Las Vegas Sun* publisher Hank Greenspun wrote in his front-page column, "Where I Stand":

> What will we tell the widow and children of a husband and father who was burned to death or suffocated through smoke inhalation when they ask the hotel owners, "Why did my husband or father have to die?"

176

How can the public officials who are charged with the safety of the public explain to the orphans of parents who died in the fire that the deaths did not have to be?

What assurance can we give to the millions of visitors who are daily enticed by our advertising and convention bureaus to come to Las Vegas for rest, relaxation and enjoyment that they, too, will not wind up as burnt corpses in our supposedly modern pleasure palaces? . . .

Dare we tell the millions of tourists who come here that many of our luxurious hotels . . . do not have automatic sprinkler systems or that the halls and rooms are not equipped with automatic smoke alarm signals? . . .

How is it possible for a blaze to break out at one end of a large building and spread like wildfire throughout a vast casino in a matter of minutes?

Was it too much to ask that in a building that houses 5,000 guests, the sprinklers be automatically controlled instead of manually operated and that the alarm system go off at the first breath of smoke or heat so guests don't have to look for a little red box on a wall to break the glass and pull the alarm by hand?

Is there anything we can do with a public officialdom that waits until the year 1975 for automating the safety of visitors when the Coconut Grove [nightclub] fire destroyed 491 lives and pointed up the need for automatic sprinklers and smoke alarm signals in public places back in 1942?

Greenspun went on to praise the courageous performance of medical and emergency workers, firefighters, and local citizenry while condemning public officials and industry leaders who, through indifference or greed, allowed the MGM tragedy to occur.

Gordon Vickery, head of the United States Fire Administration, spoke to a *Los Angeles Times* reporter soon after the MGM tragedy. Vickery said, "There are more than twelve thousand hotel and motel fires a year, causing ninety million dollars' damage and claiming an average of 160 lives. We usually lose people by one and twos, and the public doesn't pay much attention until a disaster like this comes along."

Predictably, in the aftermath of tragedy, public attention became intensely focused on hotel safety. In Kawaji, Japan, on November 21, 1980, the same day as the MGM fire, forty-four died in a blaze at the Kawaji Prince Hotel. Within two weeks, on December 4, 1980 in a fire at the Stouffer's Inn in Westchester County, New York, twenty-six died. Less than two months following the Las Vegas holocaust, six persons were killed and fifty-nine injured in a predawn fire January 17, 1981, at the twenty-two-story Inn on the Park Hotel in Toronto.

In Las Vegas, in the first grim days following the MGM tragedy, and on Thanksgiving Day, November 27, there were some God-given miracles to be thankful for: that at the time of the blaze, the vast casino was virtually deserted—a few hours earlier or a few hours later, thousands of people could have been engulfed by the flames; that it happened at a time of day when police, fire departments, and hospitals were on a shift change, and thus day- and night-shift employees were all on hand to respond to the disaster; that because of Operation Red Flag at Nellis Air Force Base nearby, extra helicopters and crews were available; and, especially significant, that there was no wind that morning. Winds would have created enormous problems in controlling the blaze and in rescuing the approximately eight thousand people in the building.

These were fortuitous elements in the overall picture. But what could be done to ensure that such a disaster could never happen again? Among newspaper publishers calling for the most advanced laws, and for conformance with those laws, was Hank Greenspun, who also pledged that the *Sun* would not tolerate any official apathy in Nevada. Three weeks after the MGM fire, a stricken Fred Benninger talked with Mike O'Callaghan, who joined the *Las Vegas Sun* as the newspaper's vice-president in 1979 after serving two terms as Nevada's governor. O'Callaghan wrote of their meeting:

> He has had many sleepless nights since the huge hotel fire. Only long days of hard work seem to satisfy Benninger's desire to find the cause. . . . His voice cracked as he talked about the loss of life and referred to his employees as a "team of loyal workers." His eyes flashed his pride as he mentioned the people who helped save lives from certain death in the inferno.
>
> Benninger has two major goals in life:
>
> First, he must find out what caused the blaze and turned it into a roaring hell. He is not satisfied with the answers he has received to date. He wants to share what he learns with hotels over the world.
>
> Secondly, he intends to rebuild the MGM Grand so it is bigger and safer than any hotel in the world.
>
> Said Benninger, "This hotel will have every safety feature available within the state of the art. The safety features will be the latest, and the state of the art will take precedence over the demands of the laws. . . ."

At the stockholders' meeting in Reno, December 17, 1980, corporate officials confirmed that when the MGM Grand reopened in the summer of 1981, it would have sprinklers and smoke-detection equipment in every room.

In the course of the following weeks, other hotels constructed under outdated codes announced upgrading of their fire-safety systems, although agreement to retrofit with sprinklers was by no means universal. The cost of retro-fitting and the disruption of business were cited as two reasons. Then, as opponents of mandatory retrofitting stood their ground through the winter, the unthinkable happened again—this time at the Las Vegas Hilton, a hotel in the process of installing sprinklers.

At eight o'clock on the evening of February 10, 1981, fire broke out on the eighth floor of the thirty-story resort and swept through the east wing of the three-wing hotel. It was a horrifying replay of the November 21 disaster, with screaming people smashing out windows or sliding down hastily tied-together sheets. Again, while firefighters fought the blaze, helicopters circled above the roof of the towering hotel, lifting escaping guests to safety. Again, the Red Cross and disaster volunteers established a command post at the Convention Center, and again, paramedics and hospital personnel tended to the injured and dazed.

It was a nightmare revisited. Smoke poured upward from floor to floor, and gasping, choking guests, many remembering what they had read of the MGM disaster, found themselves caught in the same deadly circumstances. Rescue workers and reporters speeding to the scene had the eerie thought: *We're getting awfully good at this.*

As the fire raged, flames moved up through an elevator shaft from the eighth floor and into the windows of each floor above it, spreading 150 feet in both directions down hotel corridors, charring walls, ceilings, and carpets. Enormous clouds of smoke billowed through the building, as firefighters and hotel employees raced through the hallways, searching for victims.

The Las Vegas Hilton, flagship hotel of the Hilton Hotel chain and largest resort in the western world, had guests booked into seventy-nine percent of its twenty-seven hundred rooms that chilly February night. In the first-floor showroom, dancer Juliet Prowse had just performed her first number before an opening-night audience of fifteen hundred, when the fire broke out. Hilton entertainment chief Dick Lane stepped to the microphone and made a quiet announcement: "There is a problem," he said, and asked guests to leave the building through the front door. On the ground floor, evacuation was orderly.

In the casino, security guards raced through the crowds, calling "This is the last hand—this is definitely the last hand," as thousands of people poured out into the night. Above, in the stricken east tower, there were as yet no sprinklers in the guest rooms or hallways to quench the deadly blaze. The fire was quickly recognized as the work of an arsonist; three other fires were set in the Hilton that night, and even as rescue operations were underway, Metro-police officers and German shepherds searched through the building for the firebug. Within twenty-four hours a twenty-three-year-old busboy was under arrest. Eight persons died in the blaze, and 252 were injured.

Some Nevada legislators still joined in a chorus of protests over the necessity of retrofitting fire-safety equipment in existing buildings. One, State Senator Lawrence Jacobsen of Minden, claimed it would hurt Nevada's tourist-based economy. On the other hand, two other Nevada senators, William Hernstadt of Las Vegas and Joe Neal of North Las Vegas, introduced a bill calling for hotel owners in Las Vegas, Reno, and elsewhere to equip their buildings "with an approved system of sprinklers for protection from fire."

Ironically, a bill, drafted in response to the MGM Grand fire, reached the floor of the Nevada Senate on February 11, the day after the tragedy at the Las Vegas Hilton.

On March 4, 1981, as legislative hearings on the proposed changes in the fire-safety code were in progress, and more than five weeks before the bill would come to a vote, Alvin Benedict, president and chief operating officer of the MGM Grand, announced that the hotel would install a fire-safety system employing the latest technology to monitor, detect, and react instantly to fire danger anywhere in the hotel.

Myram Borders, Las Vegas bureau chief of United Press International, reported, "Sprinklers, smoke detectors, and speakers will be located in every room. A ventilation-purging system will be capable of exchanging all the air inside the hotel complex within ten minutes.

She went on to say that the system would include "a central computer, with an identical backup computer, which will monitor thirteen hundred locations in the hotel and can activate a thousand safety functions to halt the spread of fire or smoke and assist the evacuation of occupants."

The five-page news release issued by the MGM Grand described a twenty-eight-by-forty-foot computerized control room that would be staffed twenty-four hours a day, the control center itself to be protected by a two-hour firewall.

The multimillion-dollar fire-control system, designed by

Johnson Controls of Milwaukee, Wisconsin, would cover installation of 33,500 sprinklers, including four in each guest room, alarm boxes every two hundred feet throughout the building, and eight thousand speakers.

In the meantime, advance bookings for the hotel continued unabated. Lloyd W. Boothby, MGM vice-president of sales and marketing, said on March 7, "At the moment, we have some forty-six million dollars' worth of guest-room bookings for future conventions and meetings. For the years 1982 and 1983, about fifty percent of the hotel's guest rooms are already booked for various groups. Our bookings extend through 1997, when the National Electrical Contractors Association will be here."

As spring arrived, MGM Hotels board chairman Fred Benninger was still not satisfied with official pronouncements that an electrical malfunction had caused the fire.

Arson was a distinct possibility, he told the *Sun*'s Harold Hyman, but he indicated that the hotel's internal investigation had not yet produced any conclusions.

Clark County Fire Chief Roy Parrish countered that Benninger's denial of an electrical cause was without foundation.

"We thoroughly discussed with their investigators what we consider as the origin of this fire, and it was an electrical short circuit," Parrish insisted.

On May 20, 1981, the *Wall Street Journal* said that the Clark County Fire Department's final report on the MGM Grand fire had been issued. According to the *Journal*, the department had concluded that the electrical system behind the Deli had not been grounded properly. Therefore, the wiring and its surrounding conduit overheated, eventually setting fire to combustible material around the conduit.

In addition, the report stated, other mistakes had been made in installing the electrical system, and there had been other "poor construction techniques."

It is probable that the last chapter of the MGM Grand Hotel fire will not be written for many years. From the legal, financial, medical, and sociological viewpoints, the events of November 21, 1980—just those few hours in the affairs of the tumultuous twentieth century—had rapidly assumed almost mythic proportions.

Five months after the tragedy, the Nevada legislature, by a wide margin, passed a bill mandating sprinkler systems in all hotels, motels, office buildings, and apartments higher than fifty-five feet and requiring sprinklers in showrooms and other public gathering places of more than five thousand square feet.

It was a beginning.

14 . . .

The Lion Roars Again

Winter. Spring. Summer.

As a result of the MGM Grand fire, almost four thousand men and women—dealers, musicians, dancers, stagehands, cooks, bakers, waiters, room clerks, PBX operators, bartenders, maids, porters, busboys, cashiers, gardeners, engineers, office workers, lifeguards, retail clerks—were instantly out of work. Only one category of job was virtually unaffected: security guards were kept on to make lonely patrols through the ghostly building, and some dealers were pressed into service for guard duty as well. At night, the light from their flashlights would flare briefly behind the darkened, broken windows as they made their rounds from floor to floor.

Thanksgiving week, motorists on the Strip slowed as they passed the hotel, and pedestrians paused on the sidewalk for a look. "That's it . . . that's the place," they said.

For the MGM's jobless employees, a view of the fire

scene was almost too painful to bear, and there were more pressing things to do: With Christmas coming, and a child to support, one young woman dealer lost no time in unlimbering an early talent in art. She got out her paints and went to restaurants and stores seeking work decorating their windows for the holidays.

Some workers found jobs in other hotels, some drove taxis, some moved in with relatives to make ends meet. (But before the hotel's reopening, $6.25 million in state unemployment benefits would be paid.)

Many of the *Jubilee* dancers, on the strength of their contracts in a new long-running show, had bought houses and condominiums. Now jobless, they sought places in other shows. And since dancers must stay in condition, a few days after the fire a dozen of them were going through their routines at Backstage Studio in a class conducted, as it happened, by choreographer Winston Hemsley, when the studio started to fill with smoke. "Oh my God, not again!" they cried as they scrambled for the exit. A minor electrical fire elsewhere in the one-story building was swiftly put out by firemen, and the shaken dancers resumed their practice.

For Hemsley, the dance class was a way to keep busy until his departure for France. With Donn Arden, Miss Bluebell, Tom Hansen, and Rich Rizzo, Hemsley left for Paris soon after to mount the new Lido de Paris show.

In the weeks to come, as reconstruction and cleanup began, the corner of Flamingo and the Strip took on new interest. The ironworkers were clambering around the building's new tower as before, and building materials and heavy equipment jammed the hotel's parking lots. Twenty-four hours a day, month after month, work crews labored to meet the July reopening deadline. With overtime, some construction men made a thousand dollars a week. Immediately after the fire, the corporation's vow to reopen the resort by midsummer had seemed unrealistic to some observers, but by May, Fred Benninger announced from the temporary MGM Grand offices in a nearby motel that the project was moving along on schedule, that eleven hundred hotel employees had already been called back to work, and that the disaster had not impaired future convention bookings.

The MGM Grand fire would leave lasting memories among everyone involved in the events of November 21, 1980.

Said survivor Marvin Schatzman eight months later from his home in St. Louis: "I've gotten over it pretty well, but my wife has not been able to get it out of her mind." The couple went to the Missouri state legislature to push for laws that would require all hotels and motels in that state to install inexpensive smoke detectors in their rooms. They had the support of their fire chief and a number of citizen groups, but the bill never got out of committee. The proposed legislation was blocked, said Schatzman, "because of the efforts of a very powerful hotel and motel lobby. They were afraid it would have cost them too much money."

Californians Stewart and Grace Krakover heard from friends in places all over the country in the wake of Stewart's ABC Radio interview the day of the fire. Frequent travelers, the Krakovers constantly advise people they know about what to do in the event of a hotel fire.

They now learn the exact fire-exit routes when checking into a hotel, and Grace Krakover, particularly, has become very aware of smells associated with fire. "One night I smelled something burning," she said in August 1981. "Some wire in a light fixture got loose, and it touched the plastic. My heart was beating a hundred times a second!"

In May, as mentioned earlier, the United States Judicial Panel on Multi-District Litigation in Washington, D.C., ordered that all MGM-inspired lawsuits across the country be brought to Las Vegas and consolidated under the jurisdiction of United States District Judge Louis C. Bechtle of Philadelphia. The panel said the action was necessary to prevent the duplication of discovery of evidence, avoid inconsistent pre-trial rulings, and conserve the resources of all parties concerned. Las Vegas attorney Neil Galatz, of the plaintiffs' interim committee, noted of the consolidation: "This stops the trial from becoming a circus, with hundreds of lawyers running around." Nevertheless, the eventual trial, with plaintiffs from as far away as Spain represented by lawyers as familiar as F. Lee Bailey and Melvin Belli, will rank as one of the largest and most complex litigations in history.

Fire safety would never again be taken for granted in Nevada, as older hotels hurried to update their fire-safety systems, even in advance of deadlines imposed by new laws. The retrofitting of sprinkler systems, smoke detectors, and other safety devices in older hotels was now mandated by Nevada law and by an ordinance passed by the Clark County Commission—an ordinance called by Commissioner Dave Canter and Clark County Fire Chief Roy Parrish "if not the toughest, among the toughest sets of safety guidelines enacted in the nation."

At the MGM Grand, as it neared its announced end-of-July reopening, newsmen were conducted through the refurbished resort. The centerpiece of the tour was the hotel's five-million-dollar fire-safety system, with a JC-80 central computer sealed off in a room with a two-hour firewall—a system exceeding the code and incorporating all the "state of the art" features pledged by Fred Benninger in the aftermath of the disaster.

The hotel would reopen as perhaps the safest public building in the world, a fact not lost upon the public, who were booking future reservations in record numbers at the born-again resort.

As the reopening countdown began, workers installed slot machines, food-service people stocked the pantries, and maids hurried from room to room making ready guest rooms and baths.

Jubilee had gone back into rehearsal some two months before, and more than one dancer in the 135-member cast confessed to a strange feeling as she found herself back on the familiar Ziegfeld Theatre stage.

As departments were reactivated, returning employees exulted at seeing the end of a long ordeal, an end that occurred "quietly" at noon on Wednesday, July 29, 1981, a day ahead of schedule, when the hotel opened its doors once again. This time, there was no fanfare. The MGM Grand Hotel was back in operation.

Among the elated employees was Rosalie Manganelli, an assistant head hostess, who had opened the Deli on the morning of November 21, watched the keno board explode, and escaped through the killer smoke seconds later. "It's all behind us," she told the *Sun*. "This is the day we've been waiting for."

Actor Cary Grant, the first guest to register on the night of July 28, was occupying a twentieth-floor suite. He strolled through the casino greeting guests.

As she served drinks at a casino bar where the bodies of several fire victims had been discovered, cocktail waitress Barbara Dennis said, "The hotel looks just about the same." The brand-new MGM Grand casino, from chandeliers to tables to marble statuary, did look just about the same, and it was a welcome sight to Sidney and Sadalle Siegel of Orlando, Florida, for instance, who had been regular visitors since 1973. "This is my choice hotel. It's like our home," said Siegel, heading for a crap table.

On the night of July 30, the curtain went up on *Jubilee* in the Ziegfeld Theatre, Dean Martin alternated with Mac Davis and Lonnie Shorr in the Celebrity Room, and in the movie theater, naturally, was the old Garbo film *Grand Hotel*.

Less than two weeks later, smoke from a welding accident at the hotel triggered alarms and sent firefighters racing to the scene. When sparks from a welder's torch caused insulation material to smolder, automatic alarms went off in rooms on two upper floors and guests were told via the public-address system to remain in their rooms and tune into a closed-circuit TV channel for instructions. The incident was minor, there was no evacuation, and the new fire-safety system worked exactly as it was meant to.

Those Who Died
at the MGM Grand Hotel

Name   *Hometown*

1. Karen Andrews, 36   Indianapolis, Indiana
2. José Luís Mata Alvarez, 46   Guanajuato, Mexico
3. David J. Asher, Jr., 39   Indianapolis, Indiana
4. Joe Bell, 31   Conway, Arkansas
5. David Blair, Jr., 26   Columbus, Ohio
6. Robert P. Bushell, M.D., 43   Fargo, North Dakota
7. Laura Castelazo, 33   Aguascalientes, Mexico
8. Victor M. Castelazo, 35   Aguascalientes, Mexico
9. Manuel Sierra de la Visitación, 43   Mexico City, Mexico
10. Susanna Elisabeth P. de Lobo, 35   Monterrey, Mexico
11. Margarita García de Sierra, 43   Mexico City, Mexico
12. León Galico, 75   Mexico City, Mexico
13. Sara Galico, 50   Mexico City, Mexico
14. Pablo Pedro Sierra García, 22   Mexico City, Mexico
15. William Gerbosi, 24   Western Springs, Illinois
16. Gustave N. Guidry, 58   Larose, Louisiana
17. Dellum Hanks, 48   Euclid, Ohio
18. Edward William Herring, 46   Irvine, California
19. Genese Bartlett Herring, 35   Irvine, California
20. Steven J. Holschuh, 30   Vail, Colorado
21. Edna Hoo, 39   Oceanside, New York
22. James L. Hoo, 38   Oceanside, New York
23. Angela Iadeluca, 33   Montreal, Canada
24. Raphael Iadeluca, 45   Montreal, Canada
25. Richard Oren Johnson, 41   Bloomington, Minnesota
26. Blanche E. Keller, 64   Carmel, Indiana
27. Jac E. Keller, 61   Carmel, Indiana
28. Mary Ellen Knick, 40   College Park, Maryland
29. Emil J. Knick, 49   College Park, Maryland
30. Teresa Levitt, 53   Los Angeles, California
31. Ellis C. Littmann, 69   Frontenac, Missouri
32. Roslyn E. Littmann, 63   Frontenac, Missouri
33. Dolores Mack, 46   Beaumont, Texas

34. Carol Ann Mayer, 36 — Parma, Ohio
35. Glenn Alan McCarthy, 59 — Milwaukee, Wisconsin
36. Genell Ruby McDowell, 48 — Memphis, Tennessee
37. Christine Lynn McGaughey, 33 — Santa Ana, California
38. Ethel McKinney, 69 — Seal Beach, California
39. Vincent James McKinney, 72 — Seal Beach, California
40. Elmira McQuithy, 70 — Marion, Indiana
41. John M. McQuithy, 63 — Marion, Indiana
42. Barbara E. Middleton, 39 — La Mesa, California
43. John F. Monaweck, 56 — Little Rock, Arkansas
44. Fernando Lobo Morales, 35 — Monterrey, Mexico
45. Janet Llewellyn Nilssen, 59 — Omaha, Nebraska
46. Donald Charles Nilssen, M.D., 59 — Omaha, Nebraska
47. Lori Ann Nose, 19 — Parma, Ohio
48. Diane Kay Pangburn, 23 — Des Moines, Iowa
49. Charles A. Palazzolo, 40 — Revere, Massachusetts
50. Daniel Peha, 23 — Mexico City, Mexico
51. Roberta A. Petersen, 23 — Chicago, Illinois
52. David Paul Potter, 24 — La Grange, Illinois
53. Edward Myron Rogall, 69 — St. Petersburg, Florida
54. Pearl Naomi Rogall, 71 — St. Petersburg, Florida
55. Barbara Jean Sanders, 37 — Indianapolis, Indiana
56. Catherine Ann Sanders, 23 — Vail, Colorado
57. David Francis Sanders, 39 — Indianapolis, Indiana
58. Donald M. Shaffer, 44 — Morgantown, West Virginia
59. Richard E. Sipfle, 46 — Birmingham, Michigan
60. Gary Stephens, 43 — Oklahoma City, Oklahoma
61. Sammie Lee Stephens, 43 — Oklahoma City, Oklahoma
62. Allan M. Soshnik, 31 — Atlanta, Georgia
63. Barbara Flo Soshnik, 30 — Atlanta, Georgia
64. Thomas Spagnola, 40 — Des Moines, Iowa
65. James E. Thebeault, 32 — Mansfield, Ohio
66. Diane Thompson, 34 — Glendale, California
67. Tom Thompson, 36 — Glendale, California
68. Andrés Torres, 25 — Guanajuato, Mexico
69. Charles Bufford Trammel, Jr., 53 — Belle Glade, Florida
70. Patricia Louise Tunis, 60 — North Hollywood, California
71. Allan M. Unold, 37 — North Babylon, New York
72. Rosalia Unold, 36 — North Babylon, New York
73. José Luís Vásquez, 25 — Mexico City, Mexico
74. Mary Ann Vassoughi, 41 — Lower Burrell, Pennsylvania
75. Houshang Vassoughi, M.D., 44 — Lower Burrell, Pennsylvania

THE EMPLOYEES:

Name	Department
76. John Frederick Ashton, 46	Room-service waiter
77. Elizabeth M. Barresi, 53	Maid
78. María Lucy Capetillo, 31	Room-service waitress
79. Willie Lee Duncan, 55	Maid
80. Mark William Hicks, 25	Slot mechanic
81. Joseph Odell Hudgins, 37	Security guard
82. Sherman Grant Pickett, 53	Security guard
83. Phyllis Jean Thomas, 20	Cashier
84. Clarence James White, 65	Baker

APPENDIX

Hotel Fires:
How to Survive

There are twelve thousand hotel fires in the United States every year—fires from careless smoking, fires in kitchens, fires in linen rooms, and fires in air-conditioning and electrical systems.

Most are immediately brought under control. Some are not.

The MGM Grand Hotel fire of November 21, 1980, was the second deadliest hotel blaze in United States history. The worst occurred December 7, 1946, when 119 people died in a fire in the Winecoff Hotel in Atlanta.

The worst hotel fire in history is believed to have been that on Christmas Day, 1971, at the Hotel Seoul in Seoul, South Korea. It claimed 162 victims.

Some other disastrous hotel fires of the past 40 years:

—*61 dead, La Salle Hotel, Chicago, June 5, 1946.*
—*55 dead, Gulf Hotel, Houston, September 7, 1943.*
—*47 dead, Motel Philippines, Manila, November 14, 1977.*

—45 dead, Rossiya Hotel, Moscow, February 25, 1977.
—35 dead, Terminal Hotel, Atlanta, May 16, 1938.
—34 dead, a resort hotel in Stalheim, Norway, June 23, 1959.
—29 dead, a Melbourne, Australia, hotel, August 13, 1966.
—28 dead, Pioneer International Hotel, Tucson, December 20, 1970.
—25 dead, Surfside Hotel, Atlantic City, November 18, 1963.
—24 dead, Imperial Hotel, Bangkok, April 20, 1971.
—22 dead, Roosevelt Hotel, Jacksonville, December 29, 1963.
—20 dead, Thomas Hotel, San Francisco, January 6, 1961.
—19 dead, a Seattle hotel, March 10, 1970.
—16 dead, Coates House Hotel, Kansas City, January 28, 1978.
—14 dead, Lane Hotel, Anchorage, September 12, 1966.
—13 dead, Victoria Hotel, Dunnville, Ontario, January 26, 1969.

To this list must now be added the MGM Grand Hotel, Las Vegas, November 21, 1980, with eighty-four dead. On the same day in Kawaji, Japan, in a fire at the Kawaji Prince Hotel, forty-four died. Within two weeks, on December 4, 1980, the Stouffer's Inn fire in Westchester County, New York, killed twenty-six. On January 17, 1981, a fire at the Inn on the Park, Toronto, killed six. On February 10, 1981, the Las Vegas Hilton fire killed eight. The following article was written by Captain Richard

H. Kauffman, a fire specialist with Station 75 of the Los Angeles County Fire Department. Captain Kauffman originally prepared it as a safety briefing for his wife, an American Airlines flight attendant. It has been published in several airline publications and by many corporations for the safety of their traveling personnel. The message has been published in five foreign languages and in Braille. In the interest of safety education for hotel guests, Captain Kauffman has given us permission to include it in this book.

I have been a firefighter in Los Angeles for more than fifteen years, and I have seen many people die needlessly in building fires. It's sad, because most could have saved themselves.

What you're about to read is roughly the same briefing I gave my wife on hotel safety. I do not intend to play down the dangers of hotel fires or soft soap the language. It's critical that you remember how to react, and if I shake you up a little, maybe you will.

Contrary to what you have seen on television or in the movies, fire is not likely to chase you down and burn you to death. It's the byproducts of fire that will kill you. Superheated fire gases (smoke) and panic will almost always be the cause of death long before the fire arrives, if it ever does. This is very important. You must know how to avoid smoke and panic to survive a hotel fire. With that in mind, here are a few tips.

Smoke
Where there is smoke, there is not necessarily fire.

A smoldering mattress, for instance, will produce great amounts of smoke. Air-conditioning and air-exchange systems will sometimes pick up smoke from one room and carry it to other rooms, even rooms several floors above. You should keep that in mind, because seventy percent of the hotel fires are caused by smoking and matches. In any case, your prime objective should be to leave the hotel at the first sign of smoke.

Smoke, being warmer than the surrounding air, will start accumulating at the ceiling and work its way down. The first thing you will notice is, *there are no EXIT signs!* I'll talk more about exits later; just keep in mind when you have smoke, it's too late to start looking for EXIT signs.

Another thing about smoke you should be aware of is how irritating it is to the eyes. The problem is, your eyes will take only so much irritation; then they will close. Try all you want; you won't be able to open them if there is still smoke in the area. It's one of your body's defense mechanisms.

In addition, the fresh air you want to breathe is at or near the floor. Get on your hands and knees (or belly) and *stay there* as you make your way out. Those who don't, probably won't get far.

Think about this poor man's predicament for a moment:

He wakes up at 2:30 A.M. to the smell of smoke. He puts on his pants and runs into the hallway, only to be greeted by heavy smoke. He has no idea where the exit is. He runs to the right. He's coughing and gagging, and his eyes hurt. "Where is it? *Where is it?*" Panic begins to set in.

About the time he thinks maybe he is going the wrong way, his eyes close. He can't find his way back to his room (it wasn't so bad in there). His chest hurts, and he desperately needs oxygen. Total panic sets in as he runs in the other direction.

He is completely disoriented. He cannot hold his breath any longer. We find him at 2:50—dead. What caused all the smoke? A small fire in a room where they store rollaway beds. Remember, the presence of smoke does not necessarily mean that the hotel is burning down.

Panic

Panic is defined as a sudden, overpowering terror, often affecting many people at once.

Panic is the product of your imagination running wild, and it will set in as soon as it dawns on you that you're lost or disoriented or don't know what to do. Panic is contagious, and it may spread to you. Panic is almost irreversible; once it sets in, it seems to grow. Panic will make you do things that could kill you. People in a state of panic are rarely able to save themselves.

If you understand what's going on, what to do, where to go, and how to get there, panic will not set in. The man in the example I used would not have died if he had known what to do. For instance, had he known that the exit was to the left and four doors down on the left, he could have got on his hands and knees, to a level near the floor, where there was fresh air, and started counting doorways. Even if he couldn't keep his eyes open he could feel the wall as he crawled, counting the doors. One . . . two . . . three . . .

bingo! He would *not* have panicked. He would be alive today, telling of his "great hotel fire" experience.

Exits

The elevator drops you at the twelfth floor, and you start looking for your room. "Let's see, Room 1226. Ah, here it is." You open the door and drop your luggage. *At that very moment,* turn around and go back into the hallway to check your exit. You may *never* get another chance. Don't go to the bathroom, open the drapes, plop spread-eagle on the bed, turn on the TV, or untuck your shirt. I know you're tired and want to relax, but it's absolutely essential, absolutely *critical,* that you develop the *habit* of checking your exit after you drop your luggage. It won't take thirty seconds, and believe me, *you may never get another chance.*

If there are two of you sharing a room, *both* of you locate your exit. Talk it over as you walk toward it. Is it on the left or the right? Do you have to turn a corner?

Open the exit door. What do you see? . . . Stairs or another door? (Sometimes, especially in newer hotels, there are two doors to go through.)

As you return to your room, count the doors. Is it the sixth or the seventh door? . . . I'd hate to see you crawl into a broom closet thinking it was an exit. Are you passing any rooms where your friends are staying? If there was a fire, you might want to bang on their doors as you went by. Is there anything in the hallway that would be in your way—an ice machine, maybe?

As you arrive back at your room, take one more look. Get a good mental picture of what everything looks like.

Do you think you could get to the exit with a blindfold on? This procedure takes less than one minute, and to be effective, it must become a habit. Those of you who are too lazy or tired to do it consistently are real "riverboat gamblers." There are more than twelve thousand hotel fires per year. The odds are sure to catch up with you.

Using the Exit

Should you have to leave your room quickly during the night, it's important to close the door behind you. This is very effective in keeping out fire and will minimize smoke damage to your belongings. There was a house fire in Los Angeles recently in which an entire family died. It was a three-bedroom house with a den and a family room. That night the occupants had left every door in the house open except one, and that door led to the washroom where the family dog slept.

The house, except for the washroom, was a total loss. When the fire was knocked down, firemen opened the door, to find the family dog wagging his tail. Because the door was left shut, the dog and the room were in fine shape. Some doors take hours to burn through—they are excellent fire stops. So close every door you go through. If you find smoke in the exit stairwell you can be sure people are leaving the doors open as they enter.

Always take your key with you. Get into the habit of putting the key in the same place every time you stay in a hotel. Since every hotel room has a night stand, that's an excellent location. It's close to the bed so you can grab the key when you leave without wasting time looking for it. It's important that you close your door as you leave, but it's equally important you don't lock yourself out. You may find conditions in the hallway unbearable and want to return to your room.

If you're now in the habit of checking your exit and leaving the room key on the night stand, you're pretty well prepared to leave the hotel in case of a fire, so let's "walk through it" once.

Something awakens you during the night. It could be the telephone, someone banging on the door, the smell of smoke, or some other disturbance. Whatever it is, investigate it before you go back to sleep. A popular "inn" recently had a fire, and one of the guests later said that he had been awakened by people screaming but had gone back to bed thinking it was a party. He damn near died in bed.

Let's suppose you wake up to smoke in your room. Grab your key off the night stand, roll off the bed, and head for the door on your hands and knees. Even if you can tolerate the smoke by standing, don't. You'll want to save your eyes and lungs as long as possible.

Before you open the door, feel it with the palm of one hand. If the door or knob is quite hot, don't open it. The fire could be just outside. We'll talk about what to do in that case, a little later. With the palm of your hand still on the door (in case you need to slam it shut), slowly open the door and peek into the hallway to assess conditions.

As you crawl toward the exit, stay against the wall on the side where the exit is. It's very easy to get lost or disoriented in a smoky atmosphere. If you're on the wrong side of the hallway you might crawl right on by the exit.

If you're in the middle of the hall, people who are running will trip over you. Stay on the same side as the exit, counting doors as you go.

When you reach the exit and begin to descend, it's very important that you *walk* down and hang on to the handrail as you go. Don't take this point lightly. People who will be running will knock you down, and you might not be able to get up. Just hang on and stay out of everyone's way. All you have to do now is leave the building, cross the street, and watch the action. When the fire is out and the smoke clears, you will be allowed to reenter the building. If you closed your room door when you left, your belongings should be in pretty good shape.

Smoke will sometimes get into the exit stairwell. If it's a tall building, the smoke may not rise very high before it cools and becomes heavy. This is called stacking. If your room is on the twentieth floor, for instance, you could enter the stairway and find it clear. As you descend you could encounter smoke that has stacked. Do not try to run through it—people die that way. The people running down will probably be glassy-eyed and in a panic and will knock you right out of your socks. They will run over anything in their way, including firefighters. You'll feel as though you were going upstream against the Chicago Bears, but hang on and keep heading up toward the roof.

If for some reason you try one of the doors to an upper floor and find it locked, that's normal—don't worry about it. Exit stairwells are designed so you cannot enter from the street or the roof. Once inside, however, you may *exit* at the street or the roof but cannot go from floor to floor. This is for security. When you reach the roof, prop the door open with something. This is the *only* time you will leave a door open. Any smoke in the stairwell can now vent itself to the atmosphere, and you won't be locked out.

Now find the windward side of the building (the wet-finger method is quite reliable), have a seat, and wait until they find you. Roofs have proved a safe secondary exit-and-refuge area. Stay put. Firefighters will always make a thorough search of the building looking for bodies. Live ones are nice to find.

Your Room

After you check your exit and drop the key on the night stand, there is one more thing for you to do. Become familiar with your room. See if your bathroom has a vent; all do, but some vents have electric motors. Should you decide to remain in your room, turn on the vent to help remove the smoke.

Take a good look at the window in your room. How does it open? Does it have a latch? A lock? Does it slide? Now open the window (if it opens) and look outside. What do you see? A sign? Ledges? How high up are you? Get a good mental picture of what's outside; it may come in handy. You must know how to *open* your window; and you may have to close it again.

Should you wake up to find smoke in your room and discover that the door is too hot to open or the hallway is completely charged with smoke, don't panic. Many people have defended themselves quite nicely in their rooms, and so can you. One of the first things you'll want to do is open the window to vent the smoke. I hope you learned how to open it when you checked in; it could be dark and smoky in the room. Those who didn't will probably throw a chair through the window. If there is smoke outside and you can no longer close the window, the smoke will enter your room and you will be trapped.

(The broken glass from a window can cut like a surgeon's scalpel. In one serious hotel fire, an airline captain on a layover threw a chair through the window and cut himself seriously. Don't compound your problems. Besides, if you break out of your window with a chair, you could hit a firefighter on the street below.)

If there is fresh air outside, leave the window open but keep an eye on it. At this point, most people would stay at the window, waving frantically, while their room continues to fill with smoke or the fire burns through. This procedure is not conducive to longevity. You must be aggressive and fight back.

Here are some things you can do, in any order you choose: If the room phone works, let someone know you're in there. Flip on the bathroom vent. Fill the tub with water. (Don't get in it; it's for firefighting. You'd be surprised how many people try to save themselves by getting into a tub of water. You cook lobsters and crabs in water, so you know what happens.)

Wet some sheets or towels and stuff the cracks of your door to keep it cool. Feel the walls. If they're hot, bail water on them, too. You can put your mattress up against the door and block it in place with the dresser. Keep it wet. Keep everything wet. Who cares about the mess?

A wet towel tied around your nose and mouth is an effective filter, if you fold it in a triangle, put the corner in your mouth, and breathe through your nose.

If you swing a wet towel around the room, it will help clear the smoke. If there is fire outside the window, pull down the drapes and move everything that's combustible away from the window. Bail water all around the window. Use your imagination, and you may come up with some tricks of your own. The point is, there shouldn't be any reason to panic; keep fighting until reinforcements arrive. It won't be long.

Elevators

There isn't an elevator made that can be used as a safe exit. In all fifty states, elevators, by law, cannot be considered as exits. They are complicated devices with a mind of their own. The problem is, most people know only one way out of a building—the way they came in. If it was the elevator, they're in trouble.

Elevator shafts and machinery extend through all floors of a building, and besides the shaft filling with smoke, there are hundreds of other things that could go wrong and probably will. Everyone tries to get on the elevator in an emergency. Fights break out, and people get seriously injured.

Smoke, heat, and fire do funny things to elevator call buttons, controls, and other complicated parts. A case in point: Guests in a New Orleans hotel were called on their room phones and notified of a fire on the upper floors. They were in no danger but were asked to evacuate the hotel as a precaution.

Five of the guests decided to use the elevator. It was discovered later that the elevator went down only about two floors and then, for some reason, started going up. It did not stop until it reached the fire floor. The doors came open and were held open by smoke that obscured the photo-cell light beam. Besides the five guests in the elevator who died of suffocation, firefighters noticed that every button had been pushed, probably in a frantic attempt to stop the elevator.

Elevators have killed many people, including firefighters. Several New York firefighters once used an elevator while responding to a fire on the twentieth floor of a building. They pushed eighteen, but the elevator went right on past the eighteenth floor. The doors opened on the twentieth floor to an inferno and remained open long enough to kill all the firefighters. The doors then closed, and the elevator returned to the lobby.

Hand-operated elevators are not exempt. Some elevator operators have been beaten by people fighting over the controls.

If you have any idea that there might be smoke or fire in your hotel, avoid the elevator like the plague.

Jumping

It's important that I say something about jumping, because so many people do it. Most are killed or injured in the process. I cannot tell you whether you should jump. Although there are similarities, every fire is different. I can tell you, however, what usually happens to jumpers.

If you're on the first floor, you could just open the window and climb out. From the second floor, you could probably make it with only a sprained ankle, but you must jump out far enough to clear the building. Many people hit windowsills and ledges on the way down, and they go into cartwheels. If they don't land on their heads and kill themselves, they're usually injured seriously. If you're any higher than the third floor, chances are good that you won't survive the fall. You would probably be better off fighting the fire.

Nearby buildings seem closer than they really are, and many have died trying to jump to a building that looked five feet away but was actually fifteen feet away.

Panic is what causes most people to jump. There was a fire in Brazil a few years ago in which forty people jumped from windows and all forty died. Ironically, thirty-six of those jumped after the fire was out. Many people have survived by staying put, while those around them jumped to their deaths. If you can resist panic and think clearly, you can use your own best judgment.

Calling the Fire Department

Believe it or not, most hotels will not call the fire department until they verify that there really is a fire and try to put it out themselves. Should you call the front desk to report a fire, they will almost always send the bellhop, security guard, or anyone else who's not busy to investigate. Hotels are very reluctant to disturb their guests, and fire engines in the street are quite embarrassing and tend to draw crowds.

In the New Orleans hotel fire, records show that the fire department received only one call, from a guest in one of the rooms. The desk had been notified of a fire twenty minutes earlier and had sent a security guard to investigate. His body was later found on the twelfth floor, about ten feet from the elevator.

Should you want to report a fire, or smell smoke, ask the hotel operator for an outside line for a local call. Call the fire department yourself and tell them what you smell or see and your room number in case you need to be rescued.

You needn't ever feel embarrassed about calling the fire department; that's what we're here for. We would much rather come to a small fire or smoking electrical motor that you smelled than be called twenty minutes later after six people have died. Don't let hotel "policy" intimidate you into endangering yourself. The hotel may be a little upset with you, but really—who gives a damn? The fire department will be glad you called; you may have saved many lives.

The rest is up to you. Only you can condition yourself to react in a hotel emergency. You can be well prepared by developing the habits we've talked about.

(Note: Additional copies of Captain Kauffman's complete hotel fire safety manual may be ordered for the cost of printing and handling. Send $2.50 to Jazerant Corp., 3537 Old Conejo Rd., Newbury Park, California, 91320. There are quantity discounts for organizations.)